ARBORETUM BOREALIS

A Lifeline of the Planet

Diana Beresford-Kroeger

Photographs by Christian H. Kroeger

The University of Michigan Press

Ann Arbor

Copyright © Diana Beresford-Kroeger 2010

Photographs copyright © Christian H. Kroeger

All rights reserved

Published in the United States of America by

The University of Michigan Press

Manufactured in China

⊗ Printed on acid-free paper

2013   2012   2011   2010   4   3   2   1

A CIP catalog record for this book is available from the British Library.

Frontispiece: A freshwater lake of the Boreal world

Library of Congress Cataloging-in-Publication Data

Beresford-Kroeger, Diana, 1944–
    Arboretum borealis : a lifeline of the planet / Diana Beresford-Kroeger ; photographs by Christian H. Kroeger.
       p.   cm.
    Includes bibliographical references and index.
    ISBN 978-0-472-07114-2 (cloth : alk. paper) — ISBN 978-0-472-05114-4 (pbk. : alk. paper)
    1. Taigas—Canada.   2. Taiga ecology—Canada.   3. Forest plants—Canada.   I. Kroeger, Christian H.   II. Title.
QK938.T4B47   2010
578.73'7—dc22   2010014122

THIS BOOK IS DEDICATED TO THE CHILDREN OF THE NORTH

I asked my daughter, Erika, what she wanted for her sixth birthday.
She answered with such simplicity that years later, I am still shocked
by her answer. "Mommy," she said, her young face serious with thought,
"I would like a future!"

The male catkins of the tag alder, *Alnus incana* ssp. *rugosa*

Acknowledgments

I wish to thank my husband, Christian H. Kroeger, for his ongoing encouragement to write *Arboretum Borealis*. I want to say that assisting his photography in the Boreal in the presence of black bears was not easily forgotten. Mrs. Nancy Wortman hammered away on the manuscript despite a broken leg while her husband Lynn assisted in every possible way. Stuart Bernstein, my agent, gave me the full force of his support, as did the combined talents of the University of Michigan Press. A book is never the business of one person. It only flows into print by the grace of many.

A monarch butterfly
rests before migration.

Contents

MEDICINAL PLANTS

A freshwater Boreal
stream

Introduction

Today it is necessary to see the world around us with a new vision and to understand it with new words. We have seen the light. It is the view of our home from space. Our habitat is that fragile planet Earth where the seas are launched into loops of blue and the misted land masses lie open to the sun and abreast to all the galaxies of space. We exist in this moment of divinity, together. We are bound by the past to the present as a bug is to a bean in the form of mystical mathematics, which speaks alone to our own creation, to us.

This book, *Arboretum Borealis,* began a long time ago when I was orphaned as a child. This happened in Ireland. I was made a ward of the court while decisions were being made for my future by learned judges and greedy barristers. This took three years. In the meantime the Irish folk who took me under their wings were all octogenarians. They were the Irish speakers who lived on tiny farms knuckled into the mountains that are the toenails of Ireland.

They taught me the ancient culture of their dying race amid the crowing of cocks and the clucking of laying hens. They taught me the cures and the survival tactics I would need for the future. They spoke to me of the Brehon Laws laid down by my own ancestors. They taught me the meaning of life and of love, a love that comes from the heart and encompasses nature with a passion. They told me about tears. They gave me their strength, their woolen hugs filled with the smells of milk and cream and honey. They told me that . . . I would be their last child. They repeated with great gravity, "You will be the last child of our ancient Gaelic world. The very last child. There will be no more after you."

The ideas for *Arboretum Borealis* took root shortly after I gave the keynote speech at a university in northern Ontario. I was asked to speak about forest products outside of the two-by-four mantra. I had just written *Arboretum America: A Philosophy of the Forest.* The text still rests on the tip of my tongue. I was to speak of global connections to the forest, of its miracle of medicines, of the function of the forests in the larger picture of the planet and the means to use the knowledge of the forest to fix the disasters of our days. I spoke, too, of the stewardship of the Aboriginal peoples and their covenant with nature.

On the second day the large audience was taken over by a gentleman with a microphone whose prescription for the future was to cut down half of the Boreal forest. He was most persuasive. I was in the back row, as usual enjoying the quiet comforts of medicine women to my right. In front of me loomed the head of jet black hair of a trapper chief from Hudson's Bay. I had spoken to him earlier in the day and intended to continue my conversation with him later on. Strange words seemed to pop out of his hair. I was not sure at first that I had, indeed, heard them. But the words seemed to make his head shift uncomfortably in front of me. I heard them again. I sat bolt upright. ". . . cut half of the Boreal forest." I could not believe my ears. I could see the head shift again. I peered around. Nobody, but nobody said, "Boo!" What was going on here? I could just see the multinational paper company reps seated with hands clasped in bliss with the benediction of the word *cutting*. They were in the pink. They obviously thought the audience peppered with brand-name greens and government civil servants was a pushover.

I stood up. I made my way through the waves of people. I manned the mic and said, "The circumpolar Boreal forest is the crown of the planet. It is the only inheritance of every living child born today because it is our environment. To speak of its coming down is a disgrace that is no less than a disaster. What will the world think of Canada when we will deal with the Boreal forest in exactly the same manner as the Amazon. To cut the Boreal is an act of genocidal vandalism in these days of climate change and global warming."

I went on to suggest a paid moratorium of cutting for a ten- to twenty-year span funded by the World Bank to buy time before we even contemplate such an act. I was not quite finished when my shoulder was tapped and the mic taken out of my hand. A medicine woman took to the soundwaves of the PA system. Her halting voice was filled with a regal splendor of age-old wisdom. She ended, and I directly quote, "This woman speaks with one voice with us." As I had been speaking, a semicircle around my back had been slowly forming of medicine men and women from the back row, with whom I had been so comfortable.

When she uttered those words, I have never in all my life been witness to such a silence. Nobody moved. A falling pin could have been heard. The green and the corporate world together with the government representatives turned into still, lifeless dummies. Time stood still. If a pin had dropped, it would have caused an explosion . . .

And so it was, with that strange invisible world of communication of the elders of many nations still ringing in my ears, that I set off that day into the Boreal. I was with my husband Christian. He was the photographer, while I was the lookout for the plentiful supply of black bears as he worked his cameras. During the hours of my watch the land of the Boreal itself spoke to me. The feeling was that of time married to music. I heard the notes of life. I had been well tutored in Ireland to listen. I was awash with an understanding of the meaning of all of the interconnectivity of life in that place. It was as if the song I heard was a sacred song, a prayer. It arose like a vapor as a constant from the heath at my feet. It enveloped me. I became as one with the land, the age-old land. The feelings of it stretched into a form of present that makes a newness of every second of my thoughts.

It did not end there. We boarded a ship to cross the great St. Lawrence River near the Saguenay. I was over the waters and knew that the whales were below. They were just beyond the great steel keel of the boat. They pushed forward thoughts into my mind with a firmness of communication. They came so close to me that I could feel without touch the hardness of their bodies, even their breasts. It was a feeling similar to that of every mother when she has her firstborn placed into her arms. There is a perfection to the contour and a correctness to the weight that speaks of self-identity with a message that is limbic and so deep in the subconscious mind. The whales broke that barrier. They got to that place in me. I made them a promise that I would speak for them, too, and for the importance of their water world, born so long ago from the Boreal and so dependent upon it.

So the book *Arboretum Borealis* became named for the Boreal. That name is not new. It is old. Again the Greeks claimed its fame, for Boreas was a Greek god. He was the legendary god of the north wind. Over the past two thousand years he has been personified as a fiercely bearded face blowing with all of its might from the huge mouth of the true north. So the most northerly forests on earth were called the Boreal. These forests are seated upon the planet like a monk's tonsure, trimmed, tight, and fully circumpolar. The bald spot is the ice cap of the Arctic Circle.

Geographically the Boreal is huge. The circumpolar forest makes a giant sweep across the continent of North America, taking in uneven swaths of the northern regions of Newfoundland and Labrador. It swallows up great tracts of northern Quebec, Ontario, Manitoba, Saskatchewan, Alberta, and British Columbia. It trips off the north of Alberta and steals into Nunavut, the Northwest Territories and the Yukon, ending up in Alaska. Then the Boreal alights again in Greenland going into Norway, Sweden, Russia, Finland, and northern Russia. After that it elbows into the massive land-block of Siberia and swipes the top of northern China and Korea. Finally the Boreal ends in a great swirl of the sea of Okhotsk that rounds up the islands of Sakhalin and the Kuril rosary of islands, bringing in with a great splash Hokkaido, the most northerly island of Japan. In addition, many species of the Boreal make another surprising trip. They are to be found in higher elevations of the cool temperate regions of the world with a few stray relatives found in the tropics.

The botany of the Boreal is extraordinarily diverse, making the Boreal a place of mystery and secrets. The provenance of this place is ancient. It is filled with lost languages and fossil treasures of other millennia. Even the forests of spruce keep records of past ice ages, but still hold the lichen link of when the world was young. And the soil, too, is old, spread with a black treacle of carbon decay. The streams and rivers set out their spongy growth, silting their way into silent rocks waiting for the splash of water that will make them run. The glistening mosses tell a tale of lost empires and slouch sideways waiting for a fresh return. The equisetums twist their brittle bodies to scratch the sky. A mist rises and holds this world enthralled, for everything that is of the Boreal sings a song that is already sung.

Nothing on earth compares to the Boreal in maintaining life on this planet. The circumpolar runoff from the Boreal enriches the seas with nutrients in the spring. This fires up the invisible forests of the oceans called the blue-green algae or *Cyanophyceae*. These species are the foundations of all food. The blue-green algae feed the nanoplanktons that in turn feed the krill that feed the great whales. The blue-greens sitting in their water columns, set in place by the sun, bleed almost half of the world's supply of oxygen into the atmosphere. They do this on a daily basis. They also act as a grab bag for carbon, too. They dredge the gas, carbon dioxide, into the waterworks of their world as a sea salt called carbonate.

Nothing on earth compares to the evergreens of the Boreal forests in managing the most efficient photosynthesis in the cold and on the leanest diet of light, or in acting as a passive ground coolant. Nothing compares to these forests in their ability to maintain billions, if not trillions, of tons of carbon dioxide bound in phenolic plant graves that lie on the forest floor. Nothing comes even close to the Boreal's ability to maintain the thermal gradients of the saltwater conveyor belts of the great oceans of the world. It is on these moving saline belts that our global weather patterns depend. Finally, it is from this weather that atmospheric convection drips its water as fresh water for drinking, drop by drop and season by season. Indeed it is upon these

A freshwater stream bringing marine biodiversity for whale pods and their calves

3

Introduction

global patterns of available moisture that our great civilizations of East and West have arisen. But it is on the lack of water, too, that they will quietly vanish.

Nothing on earth compares to the forests of the Boreal in acting as a clean sweep of the atmosphere to filter off the smog of pollution. The needles of the evergreens and the trichomal hairs of the deciduous trees comb the air free of this minute harmful particulate pollution. This is flushed down the tree trunks of the forest by rainwater. It is neutralized and consumed by soil organisms. Some of this pollution carries hitchhikers. These can be heavy metals or radioactive molecules from reactors or war. These are laid to eternal rest in the cold chemistry of the north. The Boreal forest regions are an atmospheric health shield for the rest of the world. These forests are a living font of biologically active medicine. The Boreal winds distribute the medicinal aerosols of antibiotics, antifungal, antiviral agents, and aseptic cleaners to purify and hydrate the air.

The biota of the Boreal invent and reinvent their own particular biodiversity. It is upon the fruits of these labors that transnational migrations depend, setting down the Boreal as a place for breeding, nesting, and feeding for the many songbird populations of the world. These populations in turn use the magnetic median of the map to navigate their southbound migration.

The Boreal evergreen forests have learned to play hardball with the sun. In this game of energy the conifers bounce the sunlight back into the atmosphere again and again until the energy simmers down to an infrared. This cooler customer is then used by the conifers to run their sugar furnaces as heat. This, in turn, keeps the forest deep green. This depth of color helps to maintain the forest functioning as a unit and stops a desertification process called the Albion Effect, a phenomenon that creates a desert caused by the blinding glare of too much reflected sun.

The circumpolar forest is becoming important because it is a forest system. The Boreal is part of the global forest. And, as every kindergarten child knows today, trees sequester carbon dioxide out of the atmosphere to make sugar. Trees also evolve oxygen. Because we live in a planetary bell jar, nothing else other than oceanic *Cyanophyceae* makes oxygen in sufficient quantities outside of the plant kingdom of trees. The more the forests come down, the more every unborn child is in danger because of a deficit of oxygen. A further loss of forests would cause a loss of atmospheric oxygen such that every child in utero would have to stay a bit more time in the womb to complete the human journey it needs to breathe independently.

In the 1950s the circumpolar Boreal forest represented 7 percent of the total global forest system. By the turn of the millennium, the year 2000, that representation had risen to 14 percent, which means that so much of the world's forest cover was being felled that the Boreal then represented nearly one-sixth of the global forest. By the year 2050, which is not too far away, the Boreal circumpolar forest will represent 21 percent of the total global forest. That is, if it is left standing at all.

Lest we forget it, the Boreal forest represents home to many nations of the world. The majority of these people are aboriginal. These aboriginal nations represent an extraordinary biodiversity. Within each nation there resides a culture and a language. Trapped within the words of their many languages are the thoughts and ideas that make up a module of philosophy that makes them unique. Out of their language and customs come myriad different approaches solving the questions of life, any one of which might well turn out to be an invaluable set of tools for our global future.

This book, *Arboretum Borealis,* gives a sampling of the biota of the Boreal forest. I have tried to relate this most northern forest to others of the global garden. I have touched on the medicines of these plants and the importance of their molecular biochemistry, wherever possible. I believe that the dynamics of molecular machines represents an open door to our combined future in sustainable living. Many ideas in this area are put forth in this book to tempt further thinking across scientific disciplines. I believe that this book will act as a bridge in this way, as *Arboretum America: A Philosophy of the Forest* has already done and is still doing.

In *Arboretum Borealis* the discussion of each tree and its

genus takes place under a repeated group of subheadings. The first of these is "The Global Garden." I believe that plant species spread out across the world a long time ago. But they are still connected, both genetically and by something unknown in the environment, be it infrasound or genetic forecasting. This is seen, very well, in the bamboo family, *Bambusa*, when over a period of many years, species of this family decide to flower. They do so all together at exactly the same time wherever they happen to be growing in the Global Garden, even though they may be separated by oceans.

A similar strange happening is associated with the higher-order seaweeds, *Fucales*. Despite the fact that the algae is being dragged and shoved by the ocean's tides on the shores of the Boreal forest system, the tiny ripe spermatozoid finds the newly formed egg cell to begin a new plant of the underwater forest. These are produced by fruiting organs called conceptacles, and the release seems to happen all over the northern world at the same time. These tiny microscopic beings travel and find one another in tides that no living being can manage.

I am also fascinated by the unique medicines each species offers, which are sometimes offered by the entire family fold of the genus itself. These I discuss under a second subheading called "Medicine." And medicine is of itself an elegant form of molecular evolution by the plant kingdom. It is of itself also, both a protection and a message. Sometimes the message is the most important aspect of the medicine. Trees are unique. Their multimillion mouths called stoma are refined to taste the air. These send out messages in a speech of chemicals that communicate on a two-way basis. Sometimes they receive and other times they send. The chemicals are aerosols that float away from the canopy into the airways around the forest. These get swept up into the prevailing winds. Fortunately the medicine of a forest leaves a long trail, and more often than not health comes in its path.

Next among the subheadings comes "Ecofunction." I coined the word *ecofunction* for another book, *Bioplanning a North Temperate Garden*, also published by the University of Michigan Press as *A Garden for Life*. Ecofunction describes the inter-connectivity with nature in the process of form and function within any plant. Plants do not live alone, divorced from nature; rather they are married to it, as a root is to the soil, as a leaf is to sunlight. There are connections in the nature of ecofunctions that transcend our understanding of the species alone. A willow tree is a good example. It is mostly a riparian species living out its life near watersheds. But the willow silently controls the body of water, too, by pumping salicylate complexes into the air, which dissolve into the water as well. The benefits are there in a thousand ways filtered to fish, to oxygenating underwater plants, to insects like moths and butterflies and to the birds that feed on them. This happens, too, with the North American native coneflower, *Echinaceae purpurea*. This plant produces an antivenom as an antidote for the local rattlesnake populations that live on their doorstep. The antivenom goes one step further and connects itself with a coffee molecule that makes the action of the antidote rapid and lifesaving. Important medicines come from these observations and subsequent analyses. They underpin our health and our ability to deal with disease.

The "Bioplan" is the next subheading I write under. This word, too, will be found in my other books. It is new. If we are to rescue nature and ourselves, we will need new words in our everyday language. *Bioplan* is one and *biodiversity* is another. The bioplan I have defined thus:

The Bioplan is the blueprint for all connectivity of life in nature. It is the fragile web which keeps each creature in balance with its neighbour. It is predation and prey. It is the victor and victim in a vast cycle of elemental life which is almost beyond our comprehension. It is the quantum mechanic of the green chloroplast without which we would all die. It is the domatal hairs on the underside of deciduous trees harbouring the parasites for aphids. It is the ultra-violet traffic light signaling system in flowers for the insect world. It is the terpene SOS produced by plants in response to invasive damage. It is the toxin trick offered by plants for the protection of butterflies. It is the mantle of man, in his life and in his death, a divine contract, to all who share this planet.

If as a society we can gather a fuller understanding of what a tree in the forest can actually do for us, then that knowledge can be used to our greater advantage. For instance, the water lover, the riparian willow, is quite often removed from the banks of streams and has disappeared from the great watersheds of the North American continent, and we wonder at the silting, at the dry riverbeds and loss of freshwater fish, over and above the pollution index. Trees can be replaced in an imaginative manner that can benefit the purse and nature. When nature is repaired, the foundations for real improvements in society are laid down to benefit not just the rich, but rich and poor alike in a stable environment. The willow is a biological filter, but it is just one of many that ensure the purity of potable water.

I have also added small discussions under a final subheading, "Design." As we move farther away from nature in our cities of concrete, design becomes more important. The tree planted in a park or in somebody else's garden that you admire becomes your tree in a way. The plant enters the landscape of your mind, and its image stays there as yours, in memory. Sometimes the trees of memory of childhood are the important ones. They can speak to us without barriers in our own personal spaces. To design is important both for the native species and their cultivars but also for the possibilities that other nonnatives might offer as solutions to the climate problems that are to come, where natives may not thrive. We have to be flexible in our thinking and in the ability to compose and orchestrate a garden. Its design shows a visual image in front of the brain like any other form of art. And art, as always, makes way for change, which is the lifeblood of any form of progress for society.

In addition to the trees of the Boreal forest system, other plant representatives of the Boreal biota are also discussed. These have been chosen for the relationship they have with the forest system. Sometimes these plants stand alone in an arena of their own importance. This sampling of species, and there are so many in the Boreal system from which a choice can be made, has been selected on purpose. Some have been chosen for their unique medicines, other for their ritualistic purpose, and others for their historical provenance in a global way. Life on the margins of the Boreal forest produces many stunning changes in plants for their own survival.

The Boreal must be looked at more closely for what it does for the planet as a whole. The close view is the more important one perhaps. The biota of the Boreal, that is the trees, the medicinal perennials and annual plants, the mosses, the lichens, and even the vast stretches of sphagnum species, produces a molecular machinery that interacts with the air. The field of knowledge around this biodynamic aspect of the molecular discharge is just in its infancy. The mass effect is unknown, as is the ability of the Boreal to both store and graveyard carbon dioxide from the atmosphere. The carbon connections of species to species are unknown, even though such transportation of mobile ground carbon does take place. The mycorrhizal adaptions, unique in themselves for such acidic and cold grounds, are not understood. And yet this forest is under threat by mining operations, oil extraction, gas pipelines, and timber merchants. It is like amputating a leg at the groin and expecting the body to be fully mobile.

The Boreal is similar to the Amazon in many ways. Both systems of forest survive on little, one because of the cold and the other because of the heat. Consequently the tree complexes for both systems live on the true margins. Nobody, but nobody, realized that the life of the tropical cloud forest is in the water vapor trapped and condensed by the cuticular layer of the evergreen trees themselves and that this system is so delicate. The soil for the Amazon is poor. It is just a medium for mechanics to hold the trees in place. Without the canopy of the trees, there is little native specie regeneration. And so the soil, too, of the Boreal is fragile. The action of the cyanophytic or bacterial element is reduced, as is the action of hydrogen ion exchange, by temperature. So growth is painfully slow. Even when this is twinned with a unique modification of the chloroplastic DNA for reduced temperature gradients, growth is still slow.

What I am really saying in this book is that the Boreal is a treasure. Happenstance, millennia ago, produced this tonsure of growth. It cannot be replaced even if we try. It is a marvel of our world and from it come our whales, the greatest mammals of the sea. The medicine man, the trapper of the north, was right in his prayer to a god unknown. We must have respect, one for another and for the world we live in. There is only one home. It belongs to us all and we will not have another. We must see the world in prayer through the eyes of our children and then our grandchildren . . . If we do that, then we will stop being blind, for true vision is of the soul. Always.

Introduction

Resin on the cones of
the black spruce,
Picea mariana

8

Nomenclature

The Latin name of all plant species, as well as the common name, is used in the text. Where specific cultivars are mentioned, the cultivar name follows the Latin species name. For example in

Salix babylonica 'Pendula',

the weeping form of the willow, *Salix babylonica* in italics denotes the species and 'Pendula', in single quotation marks, the cultivar. The genus, that is, *Salix*, always appears with the first letter in upper case, and subsequent species' descriptions appear in lower case.

Generally *Hortus Third* was used for naming, definition, and spelling of plant genus and species names.

For each genus monograph, the following convention was used: first appears the genus name, followed by the common name, followed by the family name. An example is as follows:

Genus	*Salix*
Common name	WILLOW
Family	*Saliaceae*

Soldiers help the
author with the ice
storm of 1998.

ARBORETUM BOREALIS

Caution

To all readers of *Arboretum Borealis*

The aboriginal medicines described in this book cannot be duplicated exactly outside of the Boreal region. The chemical reason for this is the worldwide use of toxic pesticides. A recent discovery has shown that many pesticides and heavy metals are carried, either in a mechanical form or in molecular aerosol form, and are distributed in a pattern dependent on the shape, size, and density of the pesticides and metals all over the globe. Another limiting factor is temperature. As climate change proceeds, especially for the north, all reaction times for chemical interaction will increase. The contamination of toxic matter is worldwide in the global garden: no region escapes, from the Himalayas to the poles. The synergy of pesticides and heavy metals both with themselves and with the medicines of the plant kingdom is generally unexplored, unknown, and therefore potentially dangerous outside of an experimental laboratory.

TREES

A young balsam fir, *Abies balsamea*. The buttress area of attachment to the trunk shows the beginning of strength to fight snow.

14

Abies

FIR

Pinaceae

THE GLOBAL GARDEN

A northern fir, *Abies balsamea,* is a descendant of an ancient tribe that stretches its family lines across the arctic north. This same tree is called the gum tree, *pīkowāhtik,* by the elders of the Cree nation of the Boreal north.

The balsam fir, *Abies balsamea,* grows in a great swath across the North American continent and ends at the borders of British Columbia. Here among the misted mountains a switch of genes takes place and an alpine fir, *A. lasiocarpa,* cuts off the swath turning upward to Alaska. In the west the touch of time has coined a biodiversity out of the fir spreading its influence afar from vale to slippery slope. These trees are, among others, the grand fir or the lowland fir, *A. grandis,* and the rugged Pacific Silver fir, *A. amabilis,* of the subalpine forests, a silvered mirror image of the splendors of the exotic east across the waters of the Pacific.

Reaching into the gemstone of Sakhalin and the Kuril Islands of Russia together with the northern island of Hokkaido, Japan, is the Sakhalin fir, *A. sachalinensis.* Here this brother of the north bends its head to early spring frosts and manages to become more tolerant of them and their killing, cell-splitting vagaries that come with the birth of the year. The flag of frost resistance is carried into the rest of the northern Boreal world by *A. sibirica,* the Russian fir. Although the leaves are densely packed on the young growing shoots of this Russian tree, they do not compare to the protective leaf density of its comrade, the Sakhalin fir.

There are fifty species of firs in the genus found in the rest of the global garden. These all grow north of the equator. Some like it hot, like the Cilician fir, *A. cilicica,* of Turkey, Syria, and Lebanon, and others believe they like it cold, like *A. spectabilis* of Nepal, Sikkim, and Bhutan, but none enjoys the refrigeration of the climate of the global garden like the balsam fir of the Boreal.

There is also a religious pop star of the fir family. It is the Sacred Fir, *A. religiosa,* of central Mexico, the habitat of the monarch butterfly, *Danaus plexippus.* This ancient species is much beloved both for its perfumed branches and its beautiful blue female cones, especially when young. Boughs of the Sacred Fir adorn mission buildings during the days and nights of religious festivals. The use of this fir is a relic of the past, a reuse of the familiar scents of more pagan times under the nose of the Pater noster as a smelling salt of the older gods.

The family of the fir has traveled all over the global garden in the past. Members of this genus can be found from sea level to elevations of over 2,500 m (7,500 ft.). The multitudes of balsam fir, *A. balsamea,* left behind a southern orphan that is closely related to it. This tree is called the southern balsam fir or the Fraser fir. And there is another nomad left behind, *A. lasiocarpa* var. *arizonica,* which rests at higher elevations in the boiling air of Arizona. This tough-skinned creature is called the cork bark fir because of the great girth of cork or suberin the bark has managed to produce for protection from the blazing sun.

In the global garden the great melting pot of time has separated the fir genus into its individual species. These species in turn have fine-tuned themselves into geographic races for survival. Some have used the north-south axis of the garden for this movement, while others have switched themselves to the height of elevation. Very little is known about the variations of race within the different species of the fir, even though such knowledge is vital for responding to the impact of climate change.

Taken together the fir family has evolved a strange mothering habit for survival of the fir children. Generally each precious seed is an elongated mass about half the breadth of a fingernail. This seed is dense. It is packed from stem to stern with a long and highly developed embryo—like mother, like child in identity. This embryo is embedded in heavy endosperm food for the nursery years. The seed with its embryo and endosperm is attached to a sail mechanism. The shape of the sail is similar to that of a Chinese sailing ship, a junk. The junk has sailed the Pacific seas for a long time. Each uneven sail with its seed is liberated from a female cone in the upper reaches of the fir tree. The sail experiences the lift generated by its fall from the height of the tree. This lift initiates flight for the seed's dispersal. But the sail also captures another kind of lift, the same kind that the Chinese junk experiences. The full flush of air pushes the sail to pass over mountains into new territories, and it seems that the higher the mountain, the greater the sail size for the little fir seed. The moral of the fir is that travel is still cheap if you are ingenious.

Of all the firs in the global garden, the balsam fir, *A. balsamea,* is the most numerous. As it quietly tends to its business of growth using carbon dioxide for food and oxygenating the Boreal area, it carries on a second campaign. This tree orchestrates insect life and fires the fuel of medicinal health into the surrounding air, the same air that sweeps south with a healing hand to touch all within the global garden.

MEDICINE

The medicine of the balsam fir, *A. balsamea,* is to be found in the resinous gum that extrudes itself from the mature bark of the tree. This is seen as clear translucent blisters. The medicine is found in the leaves and in the inner soft bark that is located just inside the rough outer bark of the trunk. There are medicinal biochemicals in the roots, but above all, the tree itself as a living organism liberates a host of aerosol biochemicals into the surrounding airways. These are accompanied by applicator chemicals that also travel in the air.

It is probably an irony that every sip of any brandy taken anywhere in the world contains a small amount of fir as an alkaloid called coniferin. This travels very rapidly in the bloodstream and is the basis of the use of brandy in resuscitation, as practiced by all St. Bernard dogs with their barrels under their chins.

For the aboriginal peoples of the Boreal forests, the fir, in all of its forms, is one of the most valuable medicinal plants. This holds true for the entire circumpolar world. The balsam fir, *A. balsamea,* is perhaps the best medicinal tree partly because its cache of pinene is of the more unusual form. It is the beta conformation that is levorotatory. This is the form more useful to the human body in its antibacterial effect. In addition there are bornyl acetate and 3-carene, which is a constituent of turpentine, together with a number of other highly volatile terpenoids; limonene is present too. This is a wetting and dispensing agent that acts as a molecular applicator to touch and holds the biochemical medicine in place on the body for healing to commence.

Global aerosols are released from the modified leaves of balsam fir. These aerosols are both antiseptic and disinfectant because of their antibiotic content. They are carried in the air with the use of bornyl acetate and limonene. The bornyl acetate is known as Malayan camphor. It functions and is identical to a molecular windmill, using the activity of the air itself to further the reach of the active pinene. The limonene helps, too, acting like a dandelion seed, albeit a molecular one, ready with a parachute to lift and float away the pinene into uncharted territory. The vast volume of pinene and its daughter compounds released into the global airways acts as an air cleanser and deodorant on a massive scale. The flush of this health-giving air affects all mammals, among which man is just one beneficiary.

The translucent resin from balsam fir was used by all northern aboriginal peoples in infusions for colds, coughs, and asthma. It was also a common cure for consumption. The inner bark was boiled and used as a tonic, as was the spring sap. Many remedies involved the use of a "sweat bath" where fresh needles were thrown on live coals. The fumes were inhaled as a treatment for colds. In other cases the bark, which held resin blisters, was dried down, then ground down to a powder. An animal fat was added and mixed to make a salve. This ointment was used for anything from arthritis to ingrown toenails to sores. In addition, a poultice was made of young shoots or dried bark that had been boiled down to a mush. It was mixed with a member of the ginseng family called wild sarsaparilla, *Aralia nudicaulis*. This preparation was used for the rapid healing of leg ulcers, cuts, scrapes, and sores.

As an aroma therapy, balsam fir is used to naturally scent creams, salves, soaps, shampoos and oils. But people who suffer from citrus allergies should avoid the use of balsam fir oil because of its high limonene content and the risk of contact dermatitis.

ECOFUNCTION

Once upon a time in the life of the world something disastrous happened. The evidence suggests that it was around 280 million years ago, a period of time was known as the Permian. Up to the Permian, the world had seen forests, gigantic stretches of virgin wilderness. This was not the wildwood we know of today. This wilderness was made up of massive club mosses, mosses, ferns, and equisetums filled with silicate and strange umbrella-like trees with strong strap-shaped leaves that nearly touched the ground. These trees were of the *Cycadales* order, and a few of them are left today, scattered throughout the warm moist tropics as reminders of another age.

Whatever happened during the time line of the Permian is still open for scientific discussion. But many facts speak for themselves. One of these was the oxygen content of the atmosphere. It was 30 percent, a richness that drove the engine of DNA to manufacture creatures of huge proportions. Then in the blink of an eye—10 million years—the atmospheric oxygen plunged to a snuffing 13 percent. This great kill-off left us with 5 percent of all species lingering on in the great oceans of the global garden and just 30 percent of land life.

From this point of destruction to today, the chloroplastic world has managed to reoxygenate the atmosphere to a blissful 21 percent. The global garden has been balanced into a harmony that has produced the placenta in mammals, where the gift of oxygen goes from the air to the mother through the placenta to the growing child safely watered in the womb.

The chloroplastic creatures that made this oxygen were the conifers, to which the fir family belongs. And the balsam fir, *Abies balsamea*, with all of its fir cousins in the global garden, is an important component in this oxygen enrichment. Somewhere in the margins of life of the Permian, when life was at its toughest point, an extraordinary thing happened. The life force of one of the ferns changed in the tropical wilderness. One ovum changed its habit. The ovum in the sexual sorus of the fern decided it needed added protection, just like the placenta in mammals. And the conifers were born, again with the first real sexual penetration and protection of the delicate sperm. So hand in hand the conifer grew with the child, and oxygen was produced for the child in silence and with the dedication of all of divinity for the benefit of life.

So the ecofunction of the balsam fir within the great stretches of circumpolar Boreal forest is to oxygenate the atmosphere of the global garden. The Boreal does this in tandem with all the other forests, be they deciduous or evergreen. But the conifers do it best of all, because they are modified to withstand extremes. They live on the cutting edge of life in the north, on mountains and in otherwise barren land.

The family of the fir, because it is on this cutting edge, produces its own antibiotic as an atmospheric aerosol for its own salvation. This product cleanses the air, and the human family picks up the benefit. This is the true value of the ecofunction of *A. balsamea* and its kin within the green tapestry of the land leading into the sea.

The balsam fir, *Abies balsamea*, moving into the western world of the alpine fir, *A. lasio-carpa*

ARBORETUM BOREALIS

Around the sacred seas of Sakhalin and the Kuril archipelago an important member of the fir stands guard. The Sakhalin fir, *A. sachalinensis*, has the most thickly packed leaves on the branches of any fir. These leaves, too, are modified underneath, but with two gray bands that carry the lines of breathing stomata for the plant. The numerous leaves offer a homegrown temperature protection to the fragile growing tips in a climate that can be capricious. The numerous fragrant leaves exhale a greater load of aerosol antibiotic into the rushing air of the Sakhalin Strait, generating a sparkling clean air for the birthing process of the greatest mammals of the oceans, the great whale mothers and their nursing calves.

The firs, especially the balsam fir, orchestrate the insects that depend on the tree with a maestro's touch. With chemical cunning, the tree produces a host of juvenile hormones, the hormones and growth regulators of the various stages of insect life. These metabolic marvels are usually produced by a small gland called the *Corpora allata* in the insect's body. They control the larval metamorphosis from birthing to breeding of insect life. But the balsam fir plays a hand in this too. The fir produces juvabiol and juvabione. These juvenile hormones help to deregulate the pattern of metamorphosis if necessary, if the tree so desires. The trees' hormones are ideal insect regulators in that they target a stage of insect development with specificity. The tree hormone has very low mammalian toxicity so it does not accumulate up the food chain to the raptors and cloven creatures. Best of all it has a low environmental persistence because its accordion molecular structure is so easily broken down by soil organisms.

In their early years, all young firs find themselves vulnerable to grazing. This is especially important in winter when the fragrant green boughs of young trees sweep the ground and create a grave temptation for rabbits and hares. Other, larger mammals, like deer, elk, caribou, and moose, with even bigger, more sensitive noses come sniffing around where the green grub lies. The fir has managed to stay one step ahead of penury by producing a nasty little chemical for its own protection. The chemical is called lasiocarpenone. It is a liver toxin, a staggering taste of death, for those who dare to nibble.

The balsam fir produces a clear edible medicinal resin that is fragrant and sweet. This resinous gum contains all the medicines of the fir rolled into one. It is topically antieczematic, rubefacient, antiseptic and has potent expectorant capabilities. This gum is used by the flying insect world of wasps and native bees to wallpaper and waterproof their homes. It is in this antibiotic and antifungal home environment that pupation takes place. Health in the insect population, like all health, is money in the bank for the Boreal and the global forests.

BIOPLAN

The balsam fir, *Abies balsamea*, does not handle the pollution of cities very well and is consequently rarely seen in the urban garden. However, the tree is a common favorite across the world as a Christmas tree because of the symmetry, its strong apical tip, its aromatic foliage, and its evergreen habit of very little leaf loss. This tree is also used for Christmas wreaths. Many an enterprising farmer or small landholder has, over the decades, found this fir to be a welcome cash crop at Christmastime. This practice will continue into the future as long as the celebration of Christmas lasts.

The gum produced by the balsam fir is an oleoresin. It has a complex chemical formula that is turning out to be very difficult to fully characterize in a laboratory. However, the resin has been used for a long time because it has strange optical properties. It has an optical transparency to sunlight similar to that of glass. Light does not bend while passing through its substance. This fact has been used in microscopy, especially in bright field microscopy at settings of the highest power. Resin is used instead of water to make a mount of the item to be examined after it has been stained. The item has the advantage of being permanent. In this way, libraries of specimens are collected for educational use. They can be used for years without loss of microscopic character, an important bonus in medicine and botany.

This oleoresin may find other interesting uses in the future. Its optical properties could possibly be used in molecular machines where the photons of sunlight could pass, to be entrapped in laser lights for quantum changes within the Bose-Einstein formulation.

Japanese consumers have a great love for the pale wood of fir. They like its watered satin texture in their homes. While the Japanese government keeps a moratorium on cutting the country's spectacular forests of the Momi fir, *A. firma,* and the Nikko fir, *A. homolepis,* there is a keen import market for this raw wood from other places in the world. The potential for this market is growing and could be fed from planted set-asides of any species of this genus.

It is an extraordinary fact that balsam fir and indeed all the Boreal firs are still being cut down for the dead-end use of pulp. The lumber is used for interior woodwork, in light structural framing, sheathing, subflooring, scaffolding, and in forms for concrete. The wood is also used in particle board, construction plywood, and container veneer, with the usual boxes and crates being made for the commercial food market.

The *Abies* of the Boreal and other forested regions are becoming very important to each nation on earth as climate change proceeds. The *Abies* protect mountainous watersheds. These trees occupy sites on elevations that are critical to the maintenance of high-quality, aerated, well-regulated streams. These streams are the progenitors of our lakes and rivers of fresh water. They carry the pattern of all of life with them. They are now more precious than they have ever been for conservation, for green space, and perhaps even for life itself.

DESIGN

The firs are species of the wild. They grow to very large proportions in their native habitats. They do not favor shallow alkaline soils, nor do they like industrial settings. The particulate pollution of industry clogs up the breathing stomata of the lower leaf surfaces, which are to be found in straight white lines. When these regions get clogged, the fir will not thrive. All fir trees bear male and female cones on the same tree. These are very beautiful and are deep purple with an undertone of red. Sometimes they appear to be violet.

There is a native fir tree local to almost every gardening zone of the global garden. Some local cultivars have very spectacular sports of reduced size and strange leaf coloring, such as the magnificent North American, the Colorado white fir, *A. concolor* 'Candicans'. This resplendent tree produces blue-gray foliage that has a silver appearance in the distance. The purple, chubby cones arise like spires out of the upper canopy, while the lower branches sweep close to the ground, almost touching it in a strikingly elegant fashion. In a country garden this tree provides both canopy and escape for the local songbird population.

There are a handful of firs that do extremely well in a dry xerophytic garden. These are species for warmer gardens because they do not favor the cold. But these firs will nonetheless add considerable summer shade in an extremely dry garden without complaint. Such trees are very useful as the climate changes into a dry summer heat and water is in short supply. One is the Greek fir, *A. cephalonica,* a species that is disease resistant and will cope with most soils, including those of an alkaline nature. Another is the Moroccan fir, *A. marocana,* whose winter resinous buds make a fine display for a dry winter-based garden. Finally, there is the Spanish mountain fir, *A. pinsapo.* This, too, is good on chalk and has been used to line streets in Spain. It is a medium tree with short, rigid green leaves that radiate outward in an attractive manner.

One of the most outstanding beautiful firs for a colder west coast garden is the Santa Lucia fir, *A. bracteata.* This distinctive tree is large. It has gorgeous brown spindle-shaped winter buds. It also bears a crop of cones whose outside bracts are elongated at each tip. This gives all of the cones a strange whiskery, drooping appearance. This mountain climber also survives on chalky ground.

Then there is the beautiful Balsam of Gilead fir, *A. balsamea,* that elbows its way into the arctic. This fragrant tree produces interesting tight, balsam-scented buds and a smart gray bark with horizontal resin canals of protruding, gleaming gum. The long cones are a fine violet-purple when young. This tree has an award-winning dwarf cultivar that is very useful for a snow-laden garden. It is *A.b.* 'Hudsonia'. The cultivar is extremely dense with a dwarfness seen also in the leaves. This is a very slow-growing shrub—which is considered to be a boon to the lazy gardener.

Fir

The conelike fruit of the tag alder, *Alnus incana* ssp. *rugosa*

Alnus

ALDER

Betulaceae

THE GLOBAL GARDEN

Long, long ago in the life of time a tree called the alder was a superstar. That was when wisdom blossomed, and cultures worldwide had a clearer view of nature. The clan of the alder, or *Alnus,* is a group of thirty close-knit members who are part of the birch or *Betulaceae* family. The *Betulaceae* family is famous for its many medicines, one of which is so important that even infants take it in a specialized toddler form. This pill is, of course, aspirin, a little biological gem that wears many hats. Aspirin is found in alder. This comes as no surprise to the First Nations of the Boreal. They are among the last of the peoples on earth with an exclusive working knowledge of this tree they affectionately call "the willow that smells," for aspirin can be volatile, too.

There are many species and massive numbers of alder still living in the circumpolar Boreal landscape. There is a little tree, *Alnus incana* ssp. *rugosa.* This is commonly called gray alder, tag alder, or hoary alder. There is a mountain alder, *A. incana* ssp. *tenuifolia,* which goes by the name of thin leaf alder. There is an attractive alder, the green alder, *A. viridis* ssp. *crispa,* sometimes also called *A. crispa,* which is a commonly used native medicinal tree. As the alder goes west into higher ground, the Canadian Sitka alder, *A. sinuata,* is a lofty species that noses its way toward the arctic. In the Siberian Boreal there is the *A. viridis* ssp. *fructicosa,* commonly called the Siberian alder. This species has spawned a number of subspecies that join company to face the frigid cold.

The remainder of the global garden bears witness to an astonishing array of alders. Some of these exist north and south of the Himalayas. Others are found in Nepal, India, and Myanmar as *A. nepalensis,* the towering tree known as the Nepalese alder. Other alders tread their way down into Peru and Colombia, while more are found in the island of Corsica offering themselves as the Italian alder, *A. cordata.* The gardens of England boast a real beauty, the black alder tree, which preens itself among the peacocks in big estates as *A. glutinosa* var. 'Pyramidalis' in a pyramid form, England's pride and joy.

The alder was revered by the ancient Celts. The druids used the word "alder," *fearn,* to denote the letter F in their script, *Ogham craobh. Ógh,* the Celtic root of *Ogham,* means whole, chaste, and inviolate. The druids must have been very aware of the power of script within a culture. Those who control the word will also control change. While the Celtic druids were plotting, the patchwork of tribes across the Irish Sea were busy, too. They were cutting logs out of their black alder, *A. glutinosa,* which grew by their carefully contained water meadows and ancient armed ponds for watering cattle. The alder logs were carved into footwear. These were heavy wooden clogs that kept their feet dry and warm and chilblains at bay. Chilblains were the pustulating sores of toes that accompanied the British winter. They followed the peasants by foot like a pet dog.

In the meantime the Germans across another fetch grew weary of their limping limbs. They cast a gimlet eye on the British footwear. Always quick to cast a line, they did a little borrowing of their own. The German adoption of alder was so complete around the feet that they added the name *elira* to the lexicon of High German for their own shoes. Then the dandified Dutch came in. Always ready with the paintbrush, they, too, made colorful commercial products of the alder, one that was worn on feet and another put on the mantle to admire.

South America was not immune to the antics of the alder. The empire of the Incas used it as a medicine and an important source of fuel. And across the world, while the Japanese were regrouping into their warrior class and were busy refining the art of the metalsmith, the silken samurai lounged under the shade of the Japanese alder, *Alnus japonica,* another striking species.

In solidarity, the First Nations of North America plodded on, passing the milestones of millennia. Their use of the alder was mostly medicinal. The dream culture of North America held the alder in its grasp and slowly leaked out the notice of its medicines. And so the pharmacopoeia of the alder became enormous, with treatments from hair loss, to the construction of magic charms for hunting, to that most useful product of female hygiene, the feminine douche.

Today it comes down to the Déné nation of the northwest and to the Algonquian peoples who inhabit the Canadian portion of the circumpolar Boreal forest. They use three species of alder as part of their pharmacopial treasury. These species have helped them to maintain their health without the pill-popping of the south. They have made extensive use of the *Alnus* species, with its unusual code of northern growth packed with painkillers.

Soon the Déné will have no home; the cutting of the Boreal will see to that. The death of a culture brings with it an end time for its knowledge. Diversity of culture still remains the only mode to plunge the mind into the wellspring of new thought. From this has always come the real body-building of history, a renaissance.

MEDICINE

For the Boreal, the medicine of the *Alnus* species is found in the female catkins called cones. These cones are used in the green juvenile stage before they have opened. The resins can be seen and felt on these cones at that stage. As they go to maturity, open, and turn brown on the limb, they are not used. Medicine is also collected from the outer bark, inner bark, whole root, and

seasonal stems. The wood of the *Alnus* has been considered to be medicinal worldwide. The more extreme the habitat for each species, the higher the yield.

One of the most important anticancer medicines yet obtained from the flora of the angiosperm kingdom is to be found in the *Alnus* of the Boreal. The remaining thirty-odd alder species in their various global positions need immediate investigation because many are becoming rare and not a few, possibly, endangered by the activities of conflict.

In general the unique aspects of the medicines of the *Alnus* are the unusual mixture of biochemicals occurring in the plant, their ability to form cis-trans three-dimensional formations for self-management, then the ratio of these in relationship to one another. In addition, chemical groups will synergize with one another. The medicines that result from such a system are strong, and the healing is rapid.

The most important of the medicines is lupeol, a triterpene. It is also a major chemotherapeutic agent. Lupeol is found in *Alnus incana* ssp. *rugosa* (syn. *A. rugosa*) of the Boreal. There is also a high ratio of tannins of around 20 percent. Betulin also occurs, as does pinosylvin. The latter is part of the resin complex of the tissues. A red dye complex can be extracted from the *Alnus*. Metal mordation changes the dye into a large variety of colors. Quercetin is found, too. This biochemical is the unique protector of the fine hemodilution function in arterioles of mammals.

Alnus incana ssp. *rugosa* was used as a spring tonic for rejuvenating the body by the Seneca. The bark was scraped from young green shoots while the resin could still be felt on them. A decoction was made of the bark. This was taken three times a day.

A decoction of the green female cones of *A. incana* ssp. *rugosa,* or one bark bundle of the fresh green bark, was used in an interchangeable manner for the treatment of the pain of venereal disease in the male. A decoction was also used to bathe this delicate tissue. A similar decoction was also used to treat sore eyes as an eyewash.

A very complex mixture of native species was used to stem

hemorrhage. These were the elm, *Ulmus*, basswood, *Tilia*, and mountain maple, *Acer spicatum*, mixed with the plants red baneberry, *Actaea rubra*, and goldenrod, *Solidago*. These were all used with *A. incana* ssp. *rugosa* to reduce the flow rate of blood and induce clotting.

A whole-root decoction of *A. incana* ssp. *rugosa* was used to treat the pain of scalds and burns. This same decoction was used to relieve the pain of menstrual cramping, as was a steam treatment of the same decoction to induce menstruation.

The most interesting use of *A. incana* ssp. *rugosa* was as a hunting charm. A strong root decoction, containing very labile hormones from betulin still present in trace amounts, was painted onto a trap. The animal, always curious about smells because smells are the traffic signals of the wild, came sniffing. The inevitable happened. Dinner was served. The charm was praised and put away for another hungry day.

Outside of the Boreal of the north, *A. incana* ssp. *rugosa* was used for diseases of the eye in a fascinating way. Four common bumblebees were caught and trapped in a box. They were kept there until they killed one another. In this way their venom level was highest. They were air-dried and ground down into a fine powder. A pinch of this powder was added to a whole-root decoction of *A. incana* ssp. *rugosa*. A piece of root a few centimeters long was used in this preparation. The dosage was 15 ml (1 tablespoon) of the powdered decoction. Two treatments were prepared and administered.

ECOFUNCTION

All over the global garden the alder has a unique relationship with fresh water. Water is a simple molecule of one oxygen and two hydrogen atoms. The hydrogens stand at a ninety-degree angle to the larger oxygen atom. That is all. That is the genius of water, its simplicity and its marriage of bonding in a duet of electron interplay that makes it flexible enough to flow, to freeze, and to form a gas. Alder grew up with water 125 million years ago. And now alder still stands by the playmate of its youth and helps to protect the fresh water of the world by acting as a ground reflux system for it by ultrafine filtration.

Without exception, alders grow where there is damp ground and running riparian habitat, by rills and rivers, by ponds and lakes, and near capes and estuaries. Some prefer rough terrain outside of the Boreal. These are the mountain-dwelling alders that again live by rills and rushing streams, the trees banked against a barrage of boulders pulling the soil back together, holding it safe from the tearing effects of rapid erosion.

The alders are seen, too, in damp, wet, sodden ground where running water is never found. The alders maintain the integrity of the landscape of the watershed of fresh water all over the world. They do it in South America in Peru and Colombia. They do it for the Boreal circumpolar forest just as they do it for the Sierra Nevada and the Baja Peninsula and they do it for the Himalayas.

The cycling and recycling of fresh water of the planet is a complex system that is not fully understood. The water vapor of the skies is related to the fresh water that we drink, as it is to the saltwater of the great oceans of the world. All of this water goes through a filtration system of gigantic proportions. This system is the lifeline of trees, the circulatory system that is composed of the open sieve tube and the angular, fully filled phloem cell with its small daughter-companion cell. These components are to be found in all conifers and in all deciduous and evergreen trees of the global forests. The process of circulation of water is called *transpiration*. This is the movement of a molecule of water from the ground through the tree and out into the atmosphere by way of the multimouths, the stomata, of the leaves. This is the system upon which all water depends. The science of it is guesswork, without a Nobel winner in sight.

The alder is part of this system of water supply. It is affected by lunar tides on land with the ebb and flow of groundwater. Its wood is like all wood, both alive and dead, showing some effects to these lunar cycles even though the deadwood is no longer part of a living whole tree. And the alder, too, can feel the effects of the coastal tides even though the connection was an eon away on the evolutionary tree of time. These things are not understood either.

In addition to the remarkable ability to live with water, the alder can do something else. The alder can fix nitrogen. The source of nitrogen is the atmosphere in the form of nitrogen dioxide and nitrogen monoxide, both gases on the list of those causing climate change. These gases dissolve out of the air into nitric and nitrous acids, both of which enter the soil. The roots of the alder associated themselves with nodular bacteria a long time ago. These bacteria play host to the alder and pass along the nitrogen to the tree. They fix it into a water-soluble fresh food for the tree. It is in this way that the alder enriches embankments along streams and rivers all over the global garden. They do this so that there is a progression of species. In the wake of the alder's death, other trees enter the ground. These trees have a greater appetite for nitrogen. This is the basis of riparian life, the great melding of water and land.

The alder benefits the watercourse around which it lives. The trees produce a chemical that is a methylated form of salicylate as a methyl ester. This chemical is commonly used by veterinarians as a counterirritant. The alder adds the methyl ester into the watercourse in extremely minute amounts. This chemical is a protective one for the scales of fish and the skins of mammals.

The alder became smart a long time ago. It arranged a stand-off with its enemies. The tree knew that it was vulnerable when it was young. The seedlings as they grew were filled with the riches of starches they had obtained from the embryonic cotyledons. These are tasty treats. So the tree for its own protection began to manufacture a nasty little chemical called pinosylvin, which is a stilbenol compound. Stilbenol is an abortive agent. The rabbits and hares who came to nibble did not stay long with this chemical warfare. They left the alder alone.

The seeds of the alder are interesting, too. They are borne in female cones called strobili. The seeds are nuts. These nuts are produced in pairs within the bracts of the strobulus. Each of the species of alder carries a different design for dispersal. Seeds that need to swim have flotation wings that also protect them in rapidly moving mountain streams. The Sitka alder, *A. sinuata*, is a good example. Others like the European alder, *A. glutinosa*, can waddle about in water for a good year without a loss of viability. The seeds in this case have a wet suit of sticky resin that is water resistant.

Damp areas by streams, rivers, and bays are staging grounds for species migration all over the global garden. The circumpolar north is one such stage for north-south migrations of songbirds and butterflies. In this arena the enriched alder produces honeydew, nectars, and pollen that are high in food value. This in turn feeds the insect population, which cycles over to birds and to butterflies such as the comma and the harvester. The harvester and its many subspecies enjoy the product of a strange microscopic life that is set on the stage of the alder, *Alnus*.

The drama takes place as follows. Ants herd their cattle called wooly aphids on the bark fields of the alder trunks. Aphids suck honeydew that pours out from underlying punctured phloem cells of the cambium skin. As the aphids increase in number, so does the production of sugar-rich honeydew. The entire ant colony benefits. A harvester butterfly comes along and lays eggs on the bark of the alder, a species to which it is specially attracted. The butterfly larvae are hidden by the wooly aphids. And because the food source is so rich, the developing butterfly larvae are ignored, and the pupation of the harvester is too short anyway for the ants' "seek and destroy" method of management. So everybody benefits in this alder niche—the ants, the aphids, and the harvester.

The ability to enrich soils in riparian areas also has an additional benefit for other surrounding species. These produce larger fruit and seeds. This food source, high in nitrogenous proteins and complex sugars that come from enrichment, feeds the migratory species with high-caliber meals. Such food spells success both for the rigors of the journey and for the reproduction that lies at the end of the trail.

This is how songbirds and many butterflies endure such marathon runs and on the way perform acts of beneficial preda-

The water world habitat of the tag alder, *Alnus incana* ssp. *rugosa*

Alder

tion and plant pollenation. Of course, the higher animals, like the deer and moose, graze on alder too. All of this is the sustainable wheel on which the machinery of the global garden runs.

BIOPLAN

As far back as 1915 the European alder, *A. glutinosa,* was used for the reforestation of the lands around St. Petersburg. In North America *Alnus* has been used to reforest coal spoil sites in Ohio. There has also been a practice of reseeding the edges of logging roads with alder so that the forest could reclaim the roadways in time. In addition, an important forestation took place in Honolulu using the Nepal alder, *A. nepalensis,* this being a 30 m (100 ft.) tree at maturity. This species was planted as set-aside forestry for timber production for future generations. Because of their nitrogen-fixing abilities, all of the alder species are important for land reclamation, especially for town sites worldwide. They represent a sustainable anchor around which riparian areas can be repaired and general erosion control can commence. Alders are short-lived trees and ideally make way for a natural forest progression after their demise.

Alder can also be used for something quite important in climate change and global warming. Because the living alder is fire-resistant, these species can be used as a firewall around areas of a forest, city, or town that are prone to burn-over in areas of high lightning strikes.

Alder species can also be used as nurse crops for the trees of the *Annonaceae* family. This tribe is also called the custard-apple family. It is a crop of the future, with weather changes on the way. Pawpaw is a cash crop for the farming community. The pawpaw fruit is delicious and has vast commercial value both for industrial food and in the cosmetic world. The vegetation is a sustainable source of pesticide manufacture also.

The familiar red-brown color of deerskin clothing comes from alder. In addition, using other mordants such as alum, a bright yellow or green-brown color can be captured from *Alnus* leaves. A fine black can be obtained from green alder, *A. viridis,* using an additional second mordant. These natural colors are to be witnessed in the quill work and birch bark ceremonial baskets of the aboriginal north.

Another very clever technique that enhanced fishing skills was used in the Canadian aboriginal north. The nets were dyed brown with alder. Underwater these nets were therefore camouflaged to mimic the river bottom debris. This alone shows that there is more to the catching of fish than just patient waiting.

Alder wood has lent itself to the art of carving for millennia in Europe. The species, as a member of the birch family, is high in resin. It can be used like pitch pine, *Pinus rigida,* as a source of pitch gum for marine caulking. This is the black pitch sealant that in the past was used on the tall-masted sailing ships and all wooden craft as a watertight sealant.

In game management, *Alnus* increases the population of the American and Eurasian woodcock. Both are spectacular birds whose numbers are on the decline.

Alnus has also been used in construction. Venice, the city of water, is largely constructed on alder posts driven into the Laguna Veneta's mud.

DESIGN

The local *Alnus* can be used to repair riparian areas. They can also be used in town sites and in housing estates where the soil has had an upheaval or in damp soggy areas. The *Alnus* can be grown as a screening hedge. Of all of the *Alnus* species, the Italian alder, *A. cordata,* the European black alder, *A. glutinosa,* and the green alder, *A. viridis,* are the most lime tolerant. The European alder, *A. glutinosa,* and speckled or gray alder, *A. incana,* have both produced two very fine garden cultivars.

The European alder, *A. glutinosa,* has a form with very acute angles in the branches that produces a strongly fastigiate shape.

This is the *A.g.* 'Pyramidalis' so useful in a damper garden. The *A. incana* has a beautiful, small, weeping tree form that was developed in Holland over a hundred years ago. It is *A.i.* 'Pendula'.

West coast gardens enjoy two beautiful alder trees. These expand into a rich canopy when they are open grown. In this setting the catkins become longer and more conspicuous. The most northern alder is the Sitka mountain alder, *A. sinuata*. Another alder of note is the Oregon red alder, *A. rubra,* whose clusters of male and female cones are conspicuous, as is the delightful canopy of deeply yellow-green leaves.

Alder

The Boreal service-
berry, *Ameliancher
alnifolia,* is a member
of the rose family.

ARBORETUM BOREALIS

Ameliancher

SERVICEBERRY

Rosaceae

THE GLOBAL GARDEN

The serviceberry is the secret tree of the North American wild-wood. The continent hosts around twenty or so species. Most are delicately elegant trees or shrubs. They are to be found in every province of Canada, and they are found east and west in the United States. There are a further two serviceberries in Asia, one eastern species and another for the west. But for some unknown botanical reason the serviceberry mainly calls North America home.

The serviceberry of the Boreal is called *Ameliancher alnifolia*. In addition to the common name of serviceberry it is also called saskatoon, Juneberry, or western shadbush. The nation of the Cree of the Boreal have a particular love of the fruit. They were probably the peoples who named it originally. They called the tree *saskwatoomina*. All of the species of *Ameliancher* of North America are medicinal trees. They are members of the rose, or *Rosaceae*, family.

The serviceberry was part of the medicinal pharmacopeia of North America. All of the aboriginal nations used its medicine and were familiar with its flowering and fruiting habits. They even told the Puritans about it when they arrived on the east coast. The Puritans did not think that they had any need for the frivolity of flowers, but they were dead wrong.

When the Puritans set foot on the shores in the seventeenth century, they brought with them a whole set of mental baggage. They had taken part in a religious revolt against the Protestant church of England and were surrounded with all the bitterness that comes with such a split. They had broken away, forever, from a church that had become too popish, with its embroidered vestments, voluptuously carved chancels, and clerical manses that would comfortably house an army.

When the Puritans had built their first little wooden chapels, they set about the business of religion in the more robust manner of the lenten monk. They reduced their worldly pleasures to a minimum, to hard benches and daily worship at scrubbed wooden altars. When the paschal full moon rolled around in later March, they ran into problems. They wanted to celebrate Easter with its agonies of flaying, crucifixion, and body piercing. To celebrate this event they needed a chaste flower. They found what they wanted in the wildwoods. It was no less than the ser-viceberry, *Ameliancher arborea*, the Puritan miracle. The tree was an answer to their prayers. So on that first Easter Sunday and every following Easter Sunday service, the faithful brought flowering boughs of *Ameliancher* into the ritual of service in their dainty chapels. They laid the snow-white blooms on the altar and inhaled the sweet fragrance of Christ's redemption of the sins of the world. And so the tree called the *Ameliancher* was christened by the Puritan faithful as the serviceberry. The new name stuck as the common name of North America today.

The downy serviceberry, *A. arborea*, of the east coast was brought over to England on the sailing ships. It was planted in Europe in 1623. From the first plantings this tree naturalized all over England and crossed the English Channel into Europe, where it still exists as an exotic species. It is to be found spreading its way pleasantly into eastern Europe.

The serviceberries of North America can all interbreed, with the result that there are many hybrids and sports across the continent. This makes the botany of the *Ameliancher* a challenge for classification and its actual life history of evolution difficult to thread out. The group can cross with the pear, *Pyrus*, and has in the past been used as a cheap root-grafting material for the more familiar pears brought into the horticultural market from Europe.

All of the serviceberries are north temperate plant species. The American grouping managed to make its way into the southern regions of the Boreal forest as the saskatoon serviceberry, *A. alnifolia*. It did this using the survival tactic of shortening the flowering and fruiting time of the tree itself. This seems to be the case, too, for its Asiatic cousins. The flowers are in a race with the emerging leaves, and this buys time for the plant to give to the fruit what it needs to develop.

A. asiatica, the Asian *Ameliancher*, is a product of eastern Asia. This large, elegant shrub with beautiful arching, flowering branches in May can be seen in China and Korea. It is found also in many places in the southern islands of Japan. But it was not tough enough to survive life on the northern island of Hokkaido. The other serviceberry of the east is the round leaf serviceberry, *A. rotundifolia*. It is found in western Asia, into the northern regions of Africa and southern Europe. This plant has larger flowers, bigger ova, and consequently larger fruit. The little round leaves have a dusting of white hair underneath to help the tree grow in the arid conditions of spring to produce a fruit crop of black, sweet fruit with the taste of blueberries.

There are few horticultural cultivars of the serviceberry worldwide. The market for crossing and interbreeding this special group of species is wide open. There is a worldwide demand for the taste of blueberries as fruit. This taste is in all of the serviceberries of the entire clan across the North American continent.

MEDICINE

The medicine of the serviceberry, *Ameliancher alnifolia*, is to be found in the roots, bark, fresh buds, and the ripe fruit, called *pommes*. In addition, serviceberry was added to other plant ingredients to make medicinal mixtures especially for the very young.

The southern species of serviceberries have been important aboriginal medicines for a very long time.

The Boreal saskatoon serviceberry, *A. alnifolia*, was part of a plant trial testing for anticancer compounds in the recent past. This species was found to have a strong antitumor component. The tests were not followed up. The species has not been biochemically characterized for its active medicinal properties. The tiny hard seeds within the pomme have cyanogenic activity, but this, too, has not been characterized.

The pomme fruit has been collected by the aboriginal nations of North America as a medicinal food. It was prepared in such a way that the cyanogenic agents were released prior to consumption. These foods are called "bush foods," and they are highly prized for their capacity to maintain enduring health. The pommes are eaten fresh and are also stored for the winter in a dry form. When dried, they are mixed with powdered meat and lard to form pemmican cakes that are called *pastiwimini-saskwatōmina*, a favorite winter dish of the Cree nation.

An old Sioux medicine man was credited with saying, "In the old days the Indians had few diseases, and so there was a demand for a small number of medicines. A medicine man usually treated one special disease at a time and treated it successfully. He did this in accordance with his dream." The serviceberry, *Ameliancher*, was one such plant of his dream.

Fresh roots of the saskatoon serviceberry, *A. alnifolia*, were harvested and boiled to make a decoction. This decoction was used for a great variety of treatments in the Boreal. It was used as an analgesic to treat pinched nerves in the back or muscular spasms, including pains of paralysis due to severe back strain. It was also used to treat coughs and colds, chest pains, and lung infections in general.

The roots were added to the stems of *A. alnifolia* for a stronger and more active decoction that was used to treat the European disease tuberculosis. In the south, an *A. arborea* stem decoction handled another, gonorrhea.

The children of the Boreal with teething fevers were treated by a root decoction of *A. alnifolia* mixed with other local herbs to reduce the nighttime temperatures of molars breaking through the gums. A fresh bud decoction of the same plant was used in the management of diarrhea.

In the southern portions of the continent the Chippewa peoples made extensive use of a very closely related species to the Boreal serviceberry, *A. alnifolia*. They used the thicket serviceberry, *A. canadensis*, root in a decoction with the roots of cherry and young oak to treat the many diseases and complaints of their womenfolk. One that was important, in which the roots were used in a steeped form, was for the prevention of miscarriage.

ECOFUNCTION

The serviceberry, *Ameliancher*, species of trees are called indicator species for the continent of North America. The buds and indeed the tips of the branches are particularly sensitive to a change in solar conditions and to the circadian rhythm that comes with a lengthening of the days. In the natural landscape they are the first trees to show signs of growth. They do so by changing color. They turn from a reddish brown to an attractive deep pink. The young stem tips are covered with white glandular hairs that are shed as growth begins. This spring growth commences despite the temperature dips of the nights. As the tree begins its flush of growth, the anthocyanin pink-red color becomes more pronounced. This triggers the fragrant white flowers to open while they are still protected by the cones of pink-colored leaves beginning to unfold.

In the wildwoods the serviceberry, along with the willow, is among the first to show its flowering face. The stamens are rich with early pollen and the nectar ripe to flow. This early food is important for the fleet of early spring native pollinators of the forest. They must be fed before they can do their work of cross-pollenation. And the serviceberry always rises to the occasion, staying in flower for ten days or so. If the weather turns wet the flowers close up and linger for a few more days to open when the sun permits.

The flowering crown of the serviceberry, *Ameliancher alnifolia*, is the first herald of a northern spring.

33

Serviceberry

The Boreal service-berry, *Ameliancher alnifolia*, is a member of the rose family.

The serviceberry rings a bell for something else that is a wonder of the riparian world. The timing is always exactly correct. When the flowers of the serviceberry begin to run with nectar, something else happens nearby in the rivers and streams that are always close to the habitat of this little tree.

The shadflies or mayflies emerge from their instar molt to the surface of the water to mate in the millions for their short day. They are like pale white moths with delicate wings. These congregate in swarming flights as massive clouds, flying several feet above the surface of the water. They of course begin the long day's night of feeding for freshwater fish. The serviceberry and the shadflies have been intertwined for such a long, long time that the serviceberry is also known by another very old name, shadbush, after the shadfly. That name is almost forgotten now.

The berry crop of the serviceberry across the Boreal is very important for the songbird population, which goes after this food with great vigor. Serviceberries are probably the first wild food to disappear from the forest canopy. They are sometimes eaten before they are ripe. In all probability the flesh of the serviceberry contains sugars in a form that help with acuity and repair of vision, like many of the blueberry, *Ericacea*, clan.

A widely ranging butterfly called the coral hairstreak, *Harkenclenus titus*, uses the foliage of the serviceberry to overwinter its eggs. These are green with a pinkish middle and become invisible in the spring growth. The eggs develop using the serviceberry as a host species.

The sweet-tasting serviceberry is on the menu for deer and rabbits. The fruit, bark, and twigs are also eaten by flying squirrels and red squirrels. Red foxes, fishers, skunks, and bears feed on the fallen berries, and beavers enjoy the winter bark in private.

BIOPLAN

The serviceberry, *Ameliancher*, is a species of the wildwoods. This tree finds the sunspots of open canopy in the forest and sets up home. The growth is very slow because the crop of flowers and pommes is produced yearly without fail. This is costly to the overall growth of the tree. But the tree in the forest has modified its growth to a lollipop form as an understory species. It is found also at the edges of forests, in clearings, by streams, and even on hillsides. And in the Boreal, the serviceberry, *A. alnifolia*, finds the dryer, sweeter ground to spread out and to grow into its multistemmed, creeping form.

The serviceberry should be bioplanned into the hedgerow system of farmers' fields for the dividends it pays in attracting early native pollinators for commercial crops. This species is an essential one for any beekeeper because the pollen is high in zinc, an essential element to increase the worker bee population in the hive. Schools, too, could benefit from having these species in schoolyards for teaching purposes. And city and urban forested areas could add the serviceberry to the understory as a series of arched walking or cycling allées. The first signs of spring would be as beneficial to a cityscape as to those creatures living in the country.

In recent years the serviceberry, *A. alnifolia*, has been bred to produce cultivars that produce a larger fruit. These fruiting trees have become popular for small city gardens. The fruits are called saskatoons. They ripen a little later than the local wild species. They are harvested in middle to late June into early July. The cultivars that are on the market could benefit from a second cross with the western Asiatic *Ameliancher*, *A. rotundifolia*, to increase again the size of the fruit and to reduce the incidence of fungal diseases to which the present cultivars are prone in potash-poor soils.

Saskatoons are an excellent fruit. They both freeze and dry well, and on cooking the flavor of blueberries rises to the palate in pies, sauces, jams, jellies, syrups, and chutneys. Saskatoons also make an excellent wine and liqueur. The fresh fruit is high in the B complex of vitamins, especially niacin. It is also high in vitamins C and A. The mineral content of the fruit seals its fate as a health food. In 100 grams of fresh fruit, there are 69 mg of calcium, 40 mg of phosphate, and 244 mg of potash and only 0.6 mg of sodium. In addition, the fruit has copper, zinc, iron, magnesium, and manganese. This makes the saskatoon a superior fruit.

In the Boreal the serviceberry stems are used as a source of shafts for arrows. They are also tied together to make the basic frame of a sweat lodge. They have been used as wooden rims in birch bark baskets. Outside of the Boreal the tough, heavy, fine-grained wood is excellent for treenware. It is also used to make small household wooden items of decoration.

The fresh wood of the serviceberry produces a chemical that has historically been used by the Boreal nations as a natural preservative for fish oil. This chemical or complex has never been characterized. There is an industrial need worldwide for natural preservatives, especially for foods, such as nuts, that carry essential oils that go rancid on contact with air over time. The serviceberry, *Ameliancher*, could fill this niche.

DESIGN

The serviceberries, *Ameliancher*, are flowering shrubs and dainty trees that are beautiful in a small garden. The downy serviceberry, *A. arborea*, thicket serviceberry, *A. canadensis*, and west coast Pacific serviceberry, *A. florida*, are beautiful slow-growing trees with a remarkable bark that looks like a sideways corduroy fabric. The flush of spring pink can be matched to the pink shades of *Narcissi* like *N.* 'Pink Charm' in a more natural garden. The fall colors are extraordinary with vivid flame colors moving into deep reds with tints of purple. The colors of the fall serviceberry are among the best of the fall palate for a north temperate garden. The colored leaves hang on the trees for a very long time and are among the last to fall with frost.

Saskatoons are excellent small trees to add to the kitchen garden. At the moment the better cultivars are all of *A. alnifolia*. These are *A.a.* 'Northline', *A.a.* 'Smokey', *A.a.* 'Honeywood', *A.a.* 'Thiessen', and *A.a.* 'Regent'. They are inclined to sucker over time. These suckers can be ground-pegged and used to propagate the tree. These fruiting trees are surprisingly long lived and frost tolerant.

The serviceberry should be part of a native garden woodland design. It should also be added to a medicinal garden. For a very tiny city garden, the interesting striped tree bark could be a feature on its own. The tree could be placed by a gate or a doorway where the soft gray lines can be noticed. The other advantage of using serviceberry is that the species performs well in semishade. The tree will orient its canopy to the sun and will enjoy cool feet in the shade.

The blood-red winter
stems of the red osier
dogwood, *Cornus
sericea*

ARBORETUM BOREALIS

Cornus

DOGWOOD

Cornaceae

THE GLOBAL GARDEN

The *Cornus* species are represented all over the northern hemisphere and parts of the southern hemisphere. There is even one in Peru. This family is made up of about forty-five shrubs and small trees. They sit in the dogwood family, which is also called the *Cornaceae* family. The name *dogwood* probably comes from a corruption of *daggerwood*, because this tightly grained wood curiously holds a sharp edge like a knife. The wood also withstands high impact. In the past, a handheld dagger brought the tree closer to its real purpose, that of being connected to the ancient ritual of the charm.

An item that is charmed is an item that holds special supernatural powers. These powers can be explained quite easily in the science of today although the discovery of these powers stretches back into other scientific cultures of the ancient past that depended on the memory of observation. A charm was spoken in the quiet stillness of prayer. The item had a sanctity of its own and was either rubbed or kept in close body contact. The ownership of a charm was part of the mental process in the landscape of the mind where it held a sentient presence much like prayer itself, for prayer alters those to whom the prayer is offered, as those who begin the process anew.

Many cultures both past and present are familiar with charms. The Cree of the present-day Boreal forest have ten different words for their local *Cornus* species, the red osier dogwood or *Cornus sericea*. This gentle red giant of a spreading shrub is used in a prayer pipe and smoked to commence the relaxation process that is always present at the generation of thoughtful prayer. This same species, too, is used to trap dreams as a dream-catcher charm. Such existential thinking is wrung out of the solitary nature of the aboriginal's world. But it is noticed all the same elsewhere, even in the depths of Wall Street.

The masters of the charm were the ancient Irish druids. In fact the words *charm* and *druid* were closely associated with magic. The word for charm in Old Irish is *cuirim faoi dhraíocht,* which means to put the person under care of the magic art of the druids. The druid priests practiced a speciality of charm laying. This divided the layman from the priest and added to the family lines of power. The druids passed their knowledge as a closely kept family secret from generation to generation—only to die from disdain or scepticism by the next religion on the map, Christianity. These charms are similar to the well-known placebo effect in modern medicine, an effect that is being revisited for its remarkable science.

The *Cornus* species as red osier dogwood, *C. sericea,* in the North American Boreal and as the Tatarian dogwood, *C. alba,* an almost identical species, in the remaining circumpolar Boreal of Siberia down into Manchuria and northern Korea, is the *Cornus* of the true Boreal forest. Outside of the north this species rises into some of the most beautiful flowering trees of the global garden. Each flower is composed of a Tudor quartet of white or greenish white bracts that announce the presence of spring from the Himalayas through China and Japan into the Kuril Islands and across the world down into Florida. These flowers are closely followed by fruits called drupes that usually contain a two-seeded bony stone that has a thin drape of flesh much like a blackberry segment.

It is this stonewalled embryo with its sometimes bland-tasting, high-energy endosperm that feeds the global flocks of migratory songbirds in their many stopping-off feeding stations. And it is this seed basket, much like storable grain, that fills the bel-

Dogwood

lies of the multitude of small mammals that inhabit the high Boreal.

The Boreal species of dogwoods drift down into the north temperate zones of the remainder of the continents of North America, Asia, and Europe. These two species retain their pattern of reduced growth even in warmer areas. They close the net for the continuity of the feeding of insects, butterflies, songbirds, and mammals, especially during migration.

MEDICINE

The medicine of the dogwood, *Cornus*, is to be found in the root bark, the skinned roots, the trunk bark, the young skinned stems and their bark, the leaves, the whole upper plant, the pith, the flowers, and the seeds. Plants grown in higher, dryer ground have a greater ratio of glycosides and their active metabolites.

The bark of the dogwoods contains an active glycoside called verbenalin as a methyl ester. It is a form that is readily metabolized by the human body. Kaempferol and the capillary protector quercetin are found together with the great emulsifying agent, ursolic acid, that has a direct and gentle action on the linings of the gut, helping with digestion.

The root tissues contain the very efficient fever-reducing complex of betulinic acid and its allies. There are also complex tannins and gallic acids. All or some of these biochemicals together act in a manner similar to the aspirin family. They mimic the effect, but their action is much more gentle and the sedative action is greater by far, especially on the parasympathetic nervous system. It is so gentle that the *Cornus* species were used in the past as a medicine for very young babies in the treatment of colic. It is said that these aboriginal medicines settled the infants down for sleep.

Outside of the Boreal north there are some *Cornus* species that have escaped the detection of science. One of these is a rare Himalayan species that exists in an unusual evergreen form. It also grows in western China. It is fragile and it likes warm air twinned with a cool, deep, rich soil. The species is called *Cornus oblonga*. The large flowers of this fragrant species are produced in the fall. The narrow evergreen leaves carry a treasury of medicine from trichomal glandular hairs. These are protected on the leaf's undersurfaces. Such a species would be expected to produce complex fever-reducing and managing medicines that, too, act on the central nervous system of man in a regenerative mode.

The dogwoods have in the past been used as efficient medicines all over the world. They have treated the young and the old. They have been used in magic, in ritual, and in prayer. The medicines of many of the aboriginal medicine men and women worldwide have been lost to us now. Only peoples of the Boreal circumpolar community together with a handful of others know of the merits of the *Cornus* species.

One exception to this is an accident of fate. In the Ozarks and in the Appalachian Mountains of the United States of America the newly minted African American population adopted the flowering dogwood, *Cornus florida*, for their dental care. Back in their own homeland of the African continent, these people used a similar species for an identical use. And so the young and slightly tougher growing tips of local *Cornus* plants were stripped of their bark. The soft central pith was rubbed against the flat region of their teeth, both back and front. The pith was then squeezed into the gum face to harden this tissue. And so the irony came about that the poorest of America had the whitest and the best teeth in the world. They were the envy of one and all, especially the lady belles of the plantation estate.

Probably the most important medicine of the dogwood is used in the treatment of ocular conditions and the implications of such use. The Chipewyan and the Cree use a strained decoction of the ripe fruit and the central pith of parenchymatous tissue to treat eyes for snow blindness and for cataracts. The species used is red osier dogwood, *C. sericea*. The Iroquois nation farther south on the continent use a different species, green osier, *C. alternifolia*, for a similar purpose. This shade-loving species is more commonly called the pagoda dogwood by horticulturalists.

A decoction of the red osier dogwood, *C. sericea*, is also used as an emetic. Bark scrapings are taken from fresh second-year wood. This decoction is also taken to increase the metabolic rate of an athlete in many common cardioactive sports.

Because the *Cornus* species have the glycoside verbenalin twinned with quercetin and betulinic acids, the decoctions of these plants have been used all over the circumpolar Boreal and elsewhere in the world for the reduction of fevers and the treatment of coughs and colds. In some instances the ripe fruit is taken to treat tuberculosis and the chest pains that come with this disease.

The magical aspects of the dogwood, especially the red osier dogwood and the Tatarian dogwood, *C. sericea* and *C. alba*, come from the glycoside verbanalin. This biochemical is highly water soluble in itself and in all of its allied forms such as verbenalol and cornic acid. The glycoside, verbanalin, functions also as an aerosol, and its somewhat bitter-sour smell can be experienced as a taste and smell simultaneously. This is the remarkable biochemical that touches the tryptophol-serotonin regulatory pathways in the body that have an action on smooth muscle relaxation and contraction. As a smoke it is inhaled deeply and sweeps into the lungs first. It travels into the body and then relaxes the mind, freeing it of the stresses and strains of the day. The recipient is left free for prayer and meditation, to rebuild the psyche again or focus the mind for important dreams.

The aboriginal world has used the favors of *Cornus* for its cornic acid content for a very long time. This biochemical is also found in the *Verbenaceae* family and its flowering trail, which accompanied it and grew in ancient and virgin forests in the south. In its action it is very similar to salicylic and acetylsalicylic acids of aspirin fame. But the effect is more gentle in action. Aspirin is currently used as an anti-inflammatory and an analgesic. It is also used as a hemodilution aid for the circulation. But aspirin also has a sidekick action that is aseptic and antifungal for the body.

In addition the dogwood has been a source of dyes, the familiar brown and black of the Boreal world. There is also a gorgeous scarlet color obtained from its southern cousin, the flowering dogwood, *C. florida*. The fresh bark from the trunk yields all of these dyes. The dyes from the dogwoods have been important continental trading items in the past for the aboriginal peoples. Some of these dyes were hoarded by women and kept as carefully as medicinal roots.

ECOFUNCTION

The dogwoods are the most numerous shrubs of the northern circumpolar forest. Their habitat is unique because they love moisture-laden soil in the full sun. In these situations the shrub grows both by sucker and tip layering into a large spreading complex that seems to be almost completely disease-resistant despite the soggy conditions of growth. Every spring brings a flowering that is visited by insects for both nectar and pollen. The fall crop of white fruit is greedily gathered. On leaf drop the red osier dogwood, *C. sericea*, turns into one of the most beautiful red colors of the wild winter landscape.

To survive the harsh conditions of northern cold, the red osier dogwood and to a lesser extent the other Boreal species have learned how to modify the biochemical content of their stems that are exposed to the elements. Red anthocyanin compounds accumulate in the epidermis for frost protection. These increase as the weather gets colder, until in the middle of winter they become a glowing beacon of scarlet red. Anthocyanin enables the process of photosynthesis to take place in such reduced light conditions as occur in a northern winter. Interestingly, a similar situation occurs in the oceans and seas all over the world with the seaweeds or algae. Those living in the intertidal areas of the lowest tides also survive in greater water depths. These, too, have a high amount of red anthocyanin complex in their cellular structure. They form the *Rhodophyta*, which is an immense class of red-colored algae species that form both the visible and deep invisible forests of the oceans. These species also photosynthesize with less light, a unique undersea adaption to depth.

The *Cornus* species of the Boreal are primarily riparian in nature. They occur at the edges of streams and small bodies of

The summer flowers
of the red osier dog-
wood, *Cornus sericea*

just as true for the underwater plant oxygenators as it is for the amphibians and fish. The *Cornus* provides a surface swab of antiseptic solution that maintains health. This health effect is multiplied as the streams flow into rivers, which join with others out to the sea.

The dogwood is used by deer as browse. As the fall produces the red multistems, these tips become the favorite treat for the passing hungry herds, who will revisit the bushes to trim them a little further down each time they pass.

In the other areas of growth of the circumpolar forest where the Tatarian dogwood, *C. alba*, exists and does identical riparian work, the local reindeer populations nose in for their treat, too. Quite often the rabbits, hares, and mice add their dental efforts to the ground-hugging branches.

The fruit crop of the dogwoods is important all over the northern world and particularly in the Boreal landscape. These species are faithful bearers. Year after year the white northern berries are formed in copious amounts and feed songbirds and larger bird populations in their magnetic passage of migration. The rich endosperm supplies quality and the flesh of the drupe itself supplies quantity. This is the pattern of feeding upon which large flocks depend for their flight and their future.

The dogwoods are important to butterflies, especially the complex genetic group called the spring azures, *Celastrina* species. These early butterflies are seen floating above frost crystals in grass looking for their host plants. The plants that have well-watered roots are their choice. In all probability the cornin complex helps to protect these most vulnerable insects in their life history. In areas of Europe such as England, the azures are becoming rare because of a change in farming practices from the small hedgerow-enclosed fields to the American style of bigger and bigger, roadside-to-roadside cropping.

Apart from the use of the red osier dogwood, *C. sericea,* and the Tatarian dogwood, *C. alba,* as whips for childish antics, these species have a place in the long history of the world. They are and have been used in aboriginal rituals. A ritual becomes part of the fabric of life over time as it gives a deeper meaning to

water and in wet or boggy areas. They are found in large numbers along the sides of rivers and lakes. Their real numbers are immense, and their allelochemical effect is great. The mother plant sees to it by chemical means that competition for trees is eliminated, keeping the low-line of the *Cornus* near the water.

The plants, as they grow in nature, produce compounds that are extremely water soluble. They are cornic acid and its complexes. These species function in a manner similar to the willow, or *Salix,* family, which produces and introduces copious amounts of salicylic acid and its allied biochemicals into the waterways. These regulatory biochemicals are antiseptic, antifungal, and antimicrobial. They protect the plant itself from disease and flush the water in the soil for sanitation. This "ultra-clean" water makes its way into our lakes and streams. The aquatic plant, fish, and mammal life systems dependent on that water all benefit from this ecoflush.

Many areas of the circumpolar Boreal forest hold bodies of fresh water that have very slow flow rates, some being completely stagnant. In these aqueous systems the threat of fungal pathogenic disease is great for the entire aquatic system. This is the process by which that life is lived. Such things have an importance. They set down the place mat at which dinner is served for humanity. They bestow order and harmony to the believer. And in doing so they deliver health, especially mental health.

BIOPLAN

The stems of the dogwood have been and still are used as canes in basketmaking. These canes are called osiers, which is an old Middle English word of the medieval period, when osier beds were bought and sold in commerce. These beds were generally the riparian willow, *Salix caprea*, of Europe. The "osier" moniker must have been transferred to the dogwoods of North America with the arrival of the Europeans. On this continent the weaving of baskets was brought to a high art form, and the baskets were especially beautiful because of the contrasts afforded by the differing native mix of red colors. The weaving itself was exceptionally complex and required great skill. There are still some artists alive who have this knowledge.

A medicinal smoking tobacco has been enjoyed using red osier dogwood, *C. sericea*. Once upon a time it was a trading item that found its way throughout the continent of North America. It was either smoked alone or mixed with another medicinal species of the heath called bearberry, *Arctostaphylos uva-ursi*. The mature leaves of bearberry were harvested just prior to the first killing frost. These were dried for smoking and mixing with the dried cambial layer of the *Cornus* stem. The mature stems of *C. sericea* were harvested some time in September. The red outside bark was carefully removed by a gentle skinning until the green, cambial, wet layer was exposed. This damp layer was scraped off and collected. It was air-dried and then rubbed into fine fragments by hand. This tissue contains the highest content of the relaxant verbanalin. This was stored for smoking.

The *Cornus* species produced mild water antiseptics. These natural biochemicals could be investigated for use in fish farms and trout farms as substitutes for antibiotic therapy now in use. They may produce less genetic trouble downstream and may be one answer to a viable alternative to the mass production of fresh- and saltwater fish protein.

DESIGN

In a northern garden the red osier dogwood, *C. sericea*, comes alive in the winter landscape. This species is excellent for a woodland garden, a wild garden, or for mass planting in a cityscape. It requires little care over and above a surface mulch of the soil to keep weeds down. It should be part of a North American medicinal garden. And it is an ideal addition to a shrub border that could contain an underplanting of small native bulbs like dogtooth violets, *Erythronium americanum*.

Interestingly, the Chinese cousin, *C. kousa*, a beautiful small tree with white flowers, is closely related to *C. sericea*. This little tree could be part of the edible landscape because it produces strawberry-like, red fruits that taste much like the custard apple of the *Annonaceae* family, with a sweet taste of pineapple mixed with mango. A little touch of selective breeding would not go astray with *C. kousa* or any other of the *Cornus* species.

The nut of the American hazel, *Corylus americana*

42

Corylus

HAZEL

Betulaceae

THE GLOBAL GARDEN

As mankind continued to chip his chert for stone warfare at the end of the Stone Age, he became very interested in the nut as a food. It is thought that the movement of a nut called the hazel or *Corylus* across the world was due almost entirely to mankind's fondness for its flavor. Indeed in North America, the Mesolithic practice of the grand savannah design came into being as primarily a nut pasture to amplify game. This was a design centered around the open solar exposure of a nut-bearing canopy, whose wild grass pasture was flash-fired in April and November. This ashing of the soil provided fertilizer, pest management, and increased herd capacity all rolled into one. The real gold came later on in the fall as a crop of nuts. These were called *pakán*, or nut, throughout the length and breadth of the continent by an aboriginal culture that knew what it was doing and why it was doing it.

Meanwhile in Europe the history of the hazel was gathering itself into a body of knowledge even before the last ice age. From pollen records of that time taken out of the peat bogs that stretch across the entire area of Europe, it appears as if the hazel as a nut tree represented as much as 75 percent of the forest canopy at one time. This canopy exists to this day in the hedgerow system of Ireland and England around extremely old fields that have been used for haymaking only. Cobb trees circle these areas of pasture. The trees are old and spreading, and the nuts are picked directly after harvest. A few decades ago they were harvested from the tops of horse-drawn hay wagons. The cobb nuts were picked as the base of the nut showed a rounded bottom of darker green peeking from the top of the involucre membrane. This mint-green color was the nut's one signal of ripening in humid air.

The hazel that blazed the food trail was almost the same hazel that occupies the Siberian Boreal and the northern tips of China, Korea, and the northern island of Japan. This is the eastern Boreal hazel, *Corylus sieboldiana* var. *mandshurica*. It is a beaked hazel very closely related to the North American Boreal species, the beaked hazel, *C. cornuta*.

Time and place have altered the naming of the hazel. The aboriginal peoples of the north used *pakán*, the general name for nut. The Greeks used the word for hooded nut, *korys*. This was followed by the Romans who used *corylus*, the general Latin name of today. The Anglo-Saxons also used the word for "hood," *haesel*, and the ancient Celts used the word *coll*. A second name entered into the annals via a Christian route. It was filbert. This new name for hazel was part of the evangelical movement of bringing nuts and berries directly to the English altar on St. Philbert's Day, which is celebrated on August 22 every year. This is the day of the nut. And indeed by this memorial day all hazels in England have been picked and have turned a fine shade of earthy brown. But if those in the High Church were carefully counting their nuts, the peasants were up to other bawdy matters. Their world revolved around the delightful occupation of pigsticking, a full-blooded November activity that involved death by sticking of an animal, either a free-range pig or a wild boar, whose belly was filled with hazelnuts. The dogs and the hunting horn accompanied this sport across Europe. This was even celebrated in art. And so the cob or cobb or cobbe name was born. *Cob* is the rounded rump of an animal in Middle English, an exact replica of the nut itself. All these names live side by side across the world today, hazel, filbert, and cobb, but they have been added to by the more anxious of society, and so Pontil nut, Lombardy and Spanish nut all grace the face of the *Corylus* species, especially that of the European nut, *C. avellana*.

A very interesting scientific situation exists for the hazel of today, and it points to a common ancestor in the ancient wildwood. All of the major species of the global garden, and there appear to be nine or so, are related. They can also be crossed. Throughout the ages they have been crossed frequently to produce an array of hybrids. But one stands out as an individual, a strikingly beautiful tree called the Turkish tree hazel, *C. colurna* var. *jacquemontii*, which is native to southeastern Europe, northern Turkey, northern Iran, the Himalayas, and China. This tree has a plastic genetic code; the chromosome number changes. This is good news for breeders all over the world because every locus of native hazel is a chance to cross and to produce a hybrid with all of its newly born charm, to compete with climate change, to act as a multiplier of food for farms, to be disease free, and above all to produce a nut flesh that competes very nicely kilo for kilo with the best of beef, organic of course, that is sold at present in the food markets.

The Boreal forest has produced a unique hazel called a beaked hazel. This covered form is found in North America and in the eastern Boreal of Russia. The species are the beaked hazel, *C. cornuta*, and the Manchu beaked hazel, *C. sieboldiana* var. *mandshurica*. The self-sterile hazel ovum, after fertilization by pollen from another hazel, grows into a nut. As the nut grows, so do the floral bracts. These move up to form a protective envelope. This keeps growing until it covers the nut body as a beaked hood, which functions like an ordinary mask. The beaked hood gives the growing nut inside its folds all the care and protection that it can muster. The hood becomes an insecticidal and fungicidal foil. The hood protects the testa, or nut shell. But the most remarkable thing that the hood gives the hazel is the ability to remain camouflaged within the tree's canopy. The long frills at the top of the nut follow the lines of the one-year-old branches. The top ornamentation holds the shade and shadows of the tree in a perfect form of invisible stillness. The tree blindsides the entire squirrel population in the global garden while it grows. This is no small feat for the kingdom of the tree!

MEDICINE

The hazel, *Corylus*, is one of six genera or groups that are to be found in the important medicinal family of the birch, *Betulaceae*. The hazel is accompanied by species of the birch, the alder, the hornbeam, and the hop hornbeam, all of which are important native medicines in the aboriginal pharmaecopia.

The *Corylus* has never been characterized scientifically. It is an open chapter waiting to be read. The medicine of the hazel nut species would be found in the beaked and nonbeaked involucre of *C. cornuta*, the beaked hazel of the Boreal, and the American hazelnut, *C. americana*. It would be found, too, in the leaf petiole sections of the leaf proper. These alone would yield, through an alcohol separation by a layer technique, a fraction of fatty chemicals that could be of great use in twenty-four-hour interior surgical bandaging that would be absorbed by the omentum tissue during recovery from surgery.

The leaves, young twigs, stems, and roots would be expected to yield chemical variations of betulin, betuloresinic acids, essential oils, saponins, betulol or the volatile sesquiterpene alcohol, apigenin dimethyl ether, betuloside, gaultherin, methyl salicylate, and ascorbic acid, or vitamin C. All of these are historically important medicinal compounds and represent a covey of important treatments worldwide.

The *Corylus* species are prone to a serious blight infection that appears to be greater in a noncoastal habitat and in specimens over twenty years old. There may well be an environmental stress factor in the presence of this infection also. The disease is called *Xanthomonas coryli*. It is a bacterial blight that is witnessed as a shrinkage in bark of the infected tree. It occurs at eye level. This blight redirects the biochemistry of the tree upon infection to produce some of the most potent biochemicals that are to be found in the arsenal of medicine. These are the taxol family, the derivatives and recombinations of which are so active in the successful treatment of cancer.

The bacteriophage components of soils surrounding infected

trees should also be examined for their medical machinery of gene transfer capacity and toxin-forming capability.

The food value of the hazelnut is first class, especially for the maintenance of health. The protein to essential fat ratio is one to five. The nuts are laden with the B complex of thiamine, riboflavin, and niacin, all good for sturdy mental health. Calcium, sodium, iron, and phosphorus with a doubling of the salts of potassium ensure that a body's biochemical machinery is well oiled.

In the practice of ritual and folklore the hazel has been used in the past for divination. These rituals are extremely ancient. The ripe nuts were burned. From the fresh smoke a medicine man or woman was able to see into the future with very special powers that were strong. In this state they were also receptive to visionary prophecy. Freshly cut, green hazel rods were also used for water divining. Hazel rods the size of a finger were used as dowsing rods primarily to find a source of water. These divining rods, which were also called dowsers, were used in locating gold, silver, and base metals. The rare individual with this ability felt an extraordinary pull for gold, less for silver, and less again for lead. The process of dowsing itself was exhausting. It was as if the person had performed a full day's work after the location of an underground spring.

The beaked hazel, *Corylus cornuta*, of the Boreal is still being used in the management of teething problems for babies and toddlers. It is also used in a separate treatment for hay fever. Throughout the continent of North America the local hazelwood, whether it be the beaked, *C. cornuta*, or the American, *C. americana*, is used with slight variations for the similar purpose of the management of fevers in young babies that quite often come with the first painful eruption of baby teeth. It would seem that this successful message of management was passed by the people of one nation to another in what is most probably a very old medicine.

The southern treatment for teething involved making a decoction of the strawberry, *Fragaria*, roots with the fresh branches of American hazel, *C. americana*. This was boiled for a half hour in 500 ml (1 pint) of well water. Two tablespoons were given to the child at intervals of three hours. After this a necklace, something like a daisy chain, of young hazel shoots, laced together, was placed on the neck of the child. It was left there until it wore off by breaking.

In the Boreal, the beaked hazel, *C. cornuta*, also was used as a necklace. Tiny young branchlets were woven together into a chain. This was placed around the infant's neck. Sometimes in addition to this a decoction of bark was taken from the hazel, the bark being scraped off in a downward manner. A quarter handful was taken and placed in 250 ml (1 cup) of boiling water, which was allowed to cool down to lukewarm. The decoction eliminated the fever.

For one treatment of hay fever the nuts were collected when they were not quite ripe, just after the first light frost. The nuts were allowed to dry down to the point where the shell was slightly loose. The nutmeats were then eaten raw. At this point the physiology of the nuts' germination potential would be different. The proteins, vitamins, and minerals would be bound. This means that the proteins would be in their least active and least water-soluble form prior to germination.

ECOFUNCTION

Nut-bearing trees are important food sources in the wild. The high protein value of the nutmeat has a cascading effect throughout the animal kingdom, beginning with the rodent population, which multiplies very rapidly with an excess of food. This is amplified up through the ranks of the cloven beasts to man, the top predator. A similar effect is seen in avian life. The hazels are eaten by larger birds, but the smaller songbirds are there, too, for the crumbs.

It is perhaps the staminous catkins that are of most importance to the insect world as a source of high-protein food. The

Infected stem of American hazel, *Corylus americana*, the trademark of medicine

45

Hazel

The male catkins, *Corylus americana*, in winter

catkins are produced in the previous growing season and elongate into the fall. But the catkins are closed and sealed with resins for the winter months. Hazelwood catkins are among the first plant organs to change with an increase of solar exposure married to a rise in temperature. And consequently they are the first also to receive frost damage. The catkins open and unfurl their anthers. The light-yellow pollen is exposed. The hazel depends on wind pollenation, so a large amount of powdery pollen is produced. This is collected by flying insects who benefit enormously from this first feast that lasts a month. This insect feeding gives a leg up to the songbirds. The length of time of production helps with the vagaries of a northern spring.

The early steel-blue hairstreak butterfly, *Erora laeta*, follows the hazel, *Corylus*. These butterflies appear with the first frosted breath of spring, flying high among the foliage of nut trees. It is possible that some betulin-type compound in the spring canopy helps in their life cycle and for these species to survive.

Hazels love a deep open sandy soil with a touch of available calcium in it. In these growing conditions the tree develops a multibranch habit and assumes an ever widening V-shape. In poorer, more acidic soils the hazel bobs and bends, putting out its most favorite branches to creep along as a propagating layer. This effort produces a huge complex of shrublike growth that is effective in erosion control and becomes an escape shrub for bird and beast. The hazel is also drought-tolerant.

BIOPLAN

The *Corylus* of the Boreal is some of the last wildwood hazel left on the planet. It is represented primarily by the two beaked hazels, *C. cornuta* and the *C. sieboldiana* var. *mandshurica*. These are both multistemmed large shrubs, but they are also closely related to the other hazels of the world, many of which are trees. They all have the potential to be crossed, with the occasional exception of the Turkish hazel, *C. colurna*, but even here the odd tree has the correct chromosome number for crossing.

These crosses would produce nuts with hybrid vigor. And, indeed, one was produced called Barcelona. It is a hybrid that, according to the members of the Northern Nut Growers Association, is mostly responsible for a fifteen-million-dollar industry in the state of Oregon alone. This industry is growing rapidly with a worldwide demand for nutmeats.

The hazelnuts of the Boreal represent a chromosome library that could be crossed to produce food trees that could be incorporated into the farming community. These could be used to produce a two-tier style of farming that would be compatible with global warming and climate change. The northern hazel chromosomes can come south for a genetic cross to produce hybrids, whereas the southern species could not readily be moved north. The circadian system in the clockwork of the chromosomes would not allow it to grow successfully in a northern habitat with shorter growing seasons.

Two-tier farming would double the food production. The drought-resistant hazels would sequester carbon dioxide out of the atmosphere and help to oxygenate at the same time. The trees would shade the fields, especially if they were planted in an east to west direction. This would help to conserve the life systems of the soil and protect them from excessive radiation. Like all other trees, they would also hold a meniscus of water in the soil from the aquifer. The function of water close to the subsurface is that the hydrogen of water can engage in hydrogen ion exchange. This is the basis of soil-to-plant traffic in food. The trees would help shade farm livestock like poultry, protecting them from melanomas. The hazelnut meats themselves would expand food diversity in local farmers' markets.

The market for the nutmeats of the hazel is complex. There is a demand for in-shell nuts that are seasonally fresh, and there is a demand for the nutmeat itself as a source of seasoning for beverages like coffee. The nutmeat market is mopped up by the candy makers, confectioners, bakers, and the jam and breakfast cereal industries. The fresh nutmeats that are still covered by their brown integument after shelling are sometimes dry roasted. They are also sold salted and nonsalted. They are roasted or fresh and often mixed with other nuts. These are packaged under nitrogen to stop the oxidation-reduction reaction of the fatty acids to prevent rancidity and preserve freshness. The hazelnut meat, either roasted or fresh, is also milled into a nut flour for the cake trade. More nuts are used by distillers in the liquor industry for nut liqueurs.

The nut tree represents a crop that can help to feed the world. It is sustainable, looking after its own business of growing while it yields a crop. The nut tree represents a new crop with an old message. It is new because most people have not shared the excitement of seeing a nut grow. It is old because deep down within all consumers is a knowledge of the ancient past for survival. It is a kind of genetic memory. And it is for no small reason that the nut trees of North America were called antifamine trees by the aboriginal peoples who lived and loved on this land.

DESIGN

The hazel, *Corylus,* is an excellent plant for a nuttery. Luckily there are many to choose from and there are some to fit into every zone of the global garden. The trees and shrubs of the *Corylus* family are extremely drought-resistant and should be planted for that reason alone. In a nuttery it is wise to also add a representative example of a local wild hazel for cross-pollenation purposes to increase the yield of the nut crop.

The species that do well in a colder garden are the beaked hazel, *C. cornuta,* the Manchu hazel, *C. sieboldiana* var. *mandshurica,* and the Siberian hazel, *C. heterophylla.* The European hazel, *C. avellana,* and the American hazel, *C. americana,* are similar in their frost-fighting capacities. The male catkins are damaged in many areas of zone 4, decreasing the nut crop, while in warmer winters both of these large shrubs will abound with a heavy crop. The Balkan tree, *C. maxima,* will survive for years in zone 4 serving as a pollinator, never producing a nut crop. The Tibetan tree, *C. tibetica,* and the Turkish tree, *C. colurna,* together with the Chinese tree, *C. chinensis,* have delightful spreading branches of elmlike leaves. All of these will grace any nuttery. The Tibetan tree produces handfuls of burrlike husks that closely resemble the European chestnut, *Castanea sativa.* For this reason this nut is also called the Spanish chestnut.

In addition to the nuttery the hazel has produced a number of interesting cultivars. Probably the most remarkable of these is the *C. avellana* 'Contorta'. This tree is also known as the corkscrew hazel or Larry Lander's Walking Stick. This tree was an accidental discovery found growing in a hedgerow in Gloucestershire in England around 1863. It has since then become an award winner and has spread itself across the world of gardening, used frequently in designs for a winter garden. The extreme and crazy contortion of the bare winter branches is delightfully contrasted by the array of long and dangling male catkins. But unfortunately this treasure is not long-lived for some reason.

Another delightful cultivar for the late summer or fall garden is the purple-leaf filbert, *C. maxima* 'Purpurea', selected from the Balkan hazel cultivars. This popular plant is also an award winner and is one of the last of the shrubs in the fall to be relieved of its canopy. The deep purple color of this hazel is identical to the canopy of the purple beech of Europe, *Fagus sylvatica* 'Purpurea'.

And then the interested gardener could do a little breeding of the hazel on the side. Using the northern warrior, the beaked hazel, *C. cornuta,* as the mother plant, a cross could be made with any of the hazels using a brown-bag technique on the insignificant flower to cross-fertilize the ovary and watch for something very different to grow. And to crow about, of course!

47

Hazel

The creeping juniper, *Juniperus horizontalis*, is the protector of the shoreline of the Boreal world.

48

Juniperus

JUNIPER

Cupressaceae

THE GLOBAL GARDEN

Gin made the juniper famous. This high-test spirit appeared out of the Protestant Dutch. Gin was born as Holland's Geneva, *jenever boom* being the Dutch word for the juniper. So the local brewmasters rolled up their sleeves and added one kilogram of fresh ripe berries of common juniper, *Juniperus communis,* to a neutralized grain mash. Each batch produced four hundred liters of clearly redefined alcohol in the form of gin. The tonic stepped up to the bar later in the game, to join gin as the universal drink, "G 'n' T."

Catgut should have made the juniper famous. But nobody wanted to remember the seamy side of life. In the modern procedure of surgery where topical embroidery is needed, catgut thread is used. Each ply of thread with its suitably curved, stainless steel needle attached to it is industrially saturated with a solution of the oil of juniper berry. This medical douche allows the catgut to hold its strength and further leads to a disinfective dissolution when it has been used as a suture.

Juniperus is the classical name given to a very widely distributed genus of some sixty species that are found growing throughout the northern hemisphere. They are under the umbrella of a very famous medicinal family called the *Cupressaceae.* The Boreal circumpolar region is literally carpeted with several species of juniper. These are all procumbent or creeping in growth and hug closely to the earth. They are found in the far north and they are found on the margins of seashores, hugging the last grain of sand close to their green bosoms, keeping it from the erosion of the tides. The common juniper is estimated to be the most widely distributed evergreen conifer in the northern regions of North America and in the remainder of the Boreal world.

There are five major species that inhabit the Boreal. They include the common juniper, *J. communis,* and the creeping juniper, *J. horizontalis.* The common juniper jumps ship to become a little more close to the ground and reappears as *J. sibirica* and shows very little change for this move. The Siberian juniper, *J. sibirica,* is the robust ground hugger of Europe, Siberia, and into the heart of Asia. Along the unprotected shores of China, Japan. and the Sakhalin region of the Kuril Islands another two important junipers creep to and hold the shoreline together. These are the shore juniper, *J. conferta,* of the islands of Japan and Sakhalin principally. The other is the Chinese creeping juniper, *J. chinensis* var. *procumbens,* which has a dwarf or *nana* form. This species inhabits the multitude of seashore regions of the islands of Japan.

In the warmer regions of the global garden the *Juniperus* rises to the height of a small tree. The red cedar, *J. virginiana,* is seen throughout the pastures of eastern and central North America. In the Rocky Mountains, the Colorado red cedar, *J. scopulorum,* shows a similar form and height. These species are also called the pencil cedars. In the Buddhist temples of upper Myanmar, the Himalayan juniper, *J. recurva* var. *coxii,* with its delightfully fragrant, orange-brown bark, is burned as a rich incense. It is thought that the sacred smoke helps prayers to rise through the sky into heaven. The fresh green boughs are used for this sacred ritual. In the Irish landscape a strange little sport poked its nose out of the rocks of Galway, Ireland, as possibly being related to the little people of legend. This small blob is called *J. communis* 'Hornibrookii'. It engulfs the form of any small object growing nearby, taking on its shape entirely in green.

Throughout history the juniper has been used by both Greek and Arab physicians in the practice of their art. There is evidence that it was grown commercially in Europe as early as the four-teenth century, when the berry was considered to be a culinary spice. It is still used as a common spice to flavor meat dishes in Sweden. The various species of juniper have been considered to be important medicines all over the world. The aboriginal peoples of the Boreal forest and those of the European and Siberian Boreal are presently the populations who make continuing use of this species.

The juniper has come into importance in England, where there is an environmental project going on in the Midlands for heath reclamation. On the list of species that are to be replanted are the junipers. These are to be massed into areas where they have previously grown. The berries in the past were eaten by game birds. The seeds of the juniper passed through the intestinal tract of the birds and the seeds were scarified by the digestive acids in the process. The etched seed coat sitting in its digestive mass, rich in bacteria, appears to be the gunpowder to fire germination. So far the conservationists have not managed to find the right mix.

The native populations of juniper are important species to conserve throughout the global garden. Their action is medicinal for mammals, insects, and man. Their ecological value is high. It will become more important with climate change. Junipers are soil protectors; they control erosion in vast areas. They control erosion in the marginal areas of life. Their effect is rarely seen and never noticed because the places where they spin their control are away from the vision of the population. They appear to be genetically dynamic, changing their needs from one place to another and forging different races while they do it. The lessons from England tell us that they might not be as easy to replace as we had originally thought.

MEDICINE

The medicine of the juniper is to be found in the immature, the fresh, and the dried berries. It is in the young branchlets obtained from the growing tips of the plants. The bark, roots, and internal wood are used in the Boreal. The wooden timbers of many species outside of the Boreal are aromatic and medicinal. In addition, dried juniper berries are extracted to form oil of juniper berry. Another oil is obtained from the fresh branches and wood of juniper; this is called oil of juniper. Both are medicinal oils.

An Asiatic juniper, the prickly juniper, *J. oxycedrus,* is steam-distilled to produce a very volatile oil called the oil of Cade. It is also called juniper tar and is used extensively in modern medicine for the treatment of various skin conditions. There is a juniper called Savin, *J. sabina,* which is found in the southern exposures of the mountains in the southern regions of Siberia that continues in a piecemeal fashion into the Caucasus. This juniper is extremely important as a stimulant veterinary drug for animals. It has a long history of use, too, but is rarely used in the human arena because of its toxicity.

All of the junipers are toxic. These species are the carriers of an extraordinary complex called Podophyllum resin. Each species and possibly each race of juniper holds a variation of this resin. Podophyllum has recently been discovered to have strong cancer-fighting drugs. These were isolated out of another toxic plant called the mayapple, mandrake root, indian apple, or the vegetable calomel. This occurs as the Himalayan mayapple, *Podophyllum hexandrum,* in the Himalayas and as *P. peltatum* all over eastern North America.

The mayapple in the past was used by the Penobscot aboriginal peoples and the Cherokee people to treat warts, deafness, and parasitic worms. But a new semisynthetic version of podophyllotoxin has been shown to have activity against herpes 1, herpes 2, influenza, and the measle virus. The drug is presently in use. It is extraordinarily powerful against testicular cancer. It is obvious that a much more extensive investigation of these effects should commence.

A very old piece of oral history regarding a medicine in North America was that of the red cedar, *J. virginiana.* This came from an extremely old Chipewyan medicine woman in the early twentieth century who had information passed to her from her great-grandmother involving the treatment of rheumatism. The small apical tip twigs of red cedar were collected. These can be harvested in North America all year round. A bundle of twigs were boiled together. The resulting decoction was sprinkled on very hot stones and the vapors inhaled or trapped or both. To trap the steam vapor a blanket was used to cover the offending limb of the patient, which was held in the steam bath. Occasionally the decoction was taken internally.

In the Boreal the leaves of the creeping juniper, *J. horizontalis*, were burned as an incense in the home. This was also combined with sweetgrass, *Hierochloe odorata*, and used as a smudge.

The Chipewyan called the common juniper, *J. communis*, the raven tree. They used as a decoction the immature green berries while they still had the pointed tip attached to the end of the berry. This was taken as a diuretic for kidney complaints.

Occasionally one such pointed, green berry of the common juniper, *J. communis*, was eaten as a cure-all. These berries were also dried down and then smoked in a pipe to treat asthma.

The inner bark of the common juniper was removed. It was placed in warm water to soften it. This was used as a bandage for open wounds. It was also used with other plants to make an infusion to treat aches and joint pains of rheumatism.

The awl-shaped leaves of the common juniper were removed when they were bright green. These were dried down. They were then powdered. This powder was used in a treatment for eczema and active psoriasis of the skin. This was also part of a secret treatment for cancer used by the medicine men and women of the north.

In addition, the fresh branches of the common juniper were added to other native species in a complex medicinal mix to produce a decoction that was used for all manner of female troubles including postpartum illness.

In the south the common juniper was used for coughs and colds, too. A small bundle of the green branches was boiled in 11.4 liters (10 quarts) of water down to 10.2 liters (9 quarts). This decoction was used as was necessary.

Ripe seeds of the common juniper, *Juniperus communis*

Again in the south, the red cedar, *J. virginiana*, was used by the Dakotas, Omahas, Poncas, and the Pawnees for colds. The twigs of the cedar were slowly burned over an open fire. The smoke was inhaled. A decoction was also used over hot stones as a vapor bath, or the steam inhaled under cover.

There is also a medicine in the female club. It is a slug of gin, a powerful uterine stimulant that helps with the perennial problem of monthly uterine cramps.

ECOFUNCTION

The war of the worlds never stops. One is the land and the other the sea. The seashore sees it all. The tides come in sometimes with enormous strength either from the lunar calendar or from rough weather. The land is always the victim fighting for every grain to keep it ashore. The junipers help in this fight all over the immense, fragile margins of land and sea. They are the common juniper, *J. communis*, the creeping juniper, *J. horizontalis*, the

Juniper

Siberian juniper, *J. sibirica,* and the two Asian shore junipers, *J. conferta* and the Chinese creeping juniper, *J. chinensis* var. *procumbens.* All of these species lie low on the shoreline. They are immune to salt spray, weather, cold, and wind. They creep along the silt and sand, plunging anchors of roots deep into a barren ground and cling to it, reclaiming it with every ounce of their green lives.

The tides have been changing recently. Once upon a time the tides were under control. These control systems stretched out into and beyond the ebb tide area. Here another jungle existed composed of species of brown algae of the *Fucales* order. These forests of brown seaweed trees fitted with paddles of leaves called lamina acted as antiturbulent agents for the shore. These jungles of forests stopped the force of the tides and conformed them backward into their own forces. The wide sticky lamina act as nurseries, too, for fish eggs and other sea life. But the fishing trawlers with their dragnets have ploughed and harrowed and felled most of these forests. The tides will come in more viciously. They will come in with greater strength and the violent storms of global warming will pack a punch with the tides to break the shoreline asunder.

So far the junipers are mostly there. However, they are not being protected. Shores are being sprayed with farmers' pesticides, and barrier areas are not great enough to save the junipers. With the seaweed forests having been diminished over the past few years, the tides have begun to take the junipers too. This invisible erosion is happening all over the shorelines of the Boreal world and in the remaining areas of the northern hemisphere. The answer is simple enough. Ban the dragnets, protect the junipers, and stop the pesticide spraying.

The ground-hugging junipers have another ecofunction, too. They protect the soil by shading it. These junipers are extraordinarily drought-resistant. The plant wears a cuticle of waxy resin that is water-repellant and heat-resistant. The juniper functions like an inanimate object on the ground, protecting the soil from the drying rays of the sun. In this manner the juniper protects the moisture levels in the soil and also the chemical character of the soil itself.

The juniper uses another technique to increase its ability to be drought-resistant. The young green leaves begin life on the plant in a prickly awl-shape. These juvenile leaves carry their stomatal breathing spaces in long white reflective bands that are glaucous. This helps the plant to retain a greater control over its own moisture needs. The more mature leaves become scalelike, and they crowd one another out. This reduces the plant's need for water in its arid habitat.

The juniper is a berry-bearing plant. It is not known at what age they begin to bear fruit. This holds true for many of the species, even for the common juniper, *J. communis.* In the spring small male and female flowers are borne on individual branchlets. The male flowers form small yellow catkins, while the female flowers do something extraordinary after fertilization. The green female flowers are composed of three to eight pointed scales. One or more of these scales become fleshy. Then they fuse together into a ball. On the surface of the ball, the etching of a melting scale can sometimes be seen. The ball is really a strobulus or cone, but in fact it looks like a berry. It hangs on the juniper for up to three years, the ripening fruit going from green to blue-black or reddish brown. It is usually covered by a white, waxy bloom.

The seeds inside the juniper berry match the plant's habitat. If the soil is poor, dry, and rough, the seeds are wrinkled and worried looking. If the soil is richer, as in the case of the ash juniper, *J. ashei,* of Missouri, the seeds have no wrinkles. This is not the case for the common juniper, *J. communis.* The seeds have grooves like grains of sand.

The seeds of all the junipers have a mind of their own. They have dormant embryos and impermeable seed coats. They have inhibitors to germination either in the seed coat or in the outer fleshy portion of the berry. They do not respond well to pre-germination tricks. They are matched perfectly to the intestinal flora of the large birds that eat them. The seeds will go through

The northern seas fed
by the lean nutrients
from the Boreal forest
system

53

Juniper

the digestive tract of a raven or large game bird and come out smiling in the acid-rich silt and pop up as an epigeal germinator with the plumule, or baby seed head, coming heads-up first. It would appear that the big birds need the juniper in its ecofunction just as much as the juniper needs the birds.

The toxin treasure trove that the junipers use to protect themselves in health is also the toxic mix that stops grazing, especially in the low-slung shrubs. The higher bushes of juniper and the juniper trees also carry an antifeeding toxin in other mixtures of podophyllic acids, podophyllotoxin, and resin.

The olive hairstreak butterfly, *Mitoura gryneus*, the loki, *M. loki*, and the juniper hairstreak, *M. siva*, depend on the juniper as a host plant for their chemical protection.

BIOPLAN

Junipers should be bioplanned around nurseries, daycares, schools, retirement homes, and hospitals. Gardens with these species in part of the design of green space would benefit the young, the old, and the sick because the low-growing juniper, especially the common juniper, *J. communis*, and its cultivars produce podophyllotoxin resins. These are topically antiviral compounds. The resin and its various acids gas off on warm summer days and get added as aerosols into the local airways. An antiviral aerosol that is naturally produced is always beneficial for the health of the young and the aging.

The various races of the common juniper and the creeping juniper of the circumpolar Boreal region could be identified. These variants could be added into the horticultural listings of these species as health-giving shrubs. These would act as smudges in warm weather to help those who were suffering from allergies and asthma or other breathing problems. These junipers would have superior berries for commercial production.

The wood of juniper has been used commercially to make oils and perfumes. The wood is used for cabinet work and for small items such as pencils. There is an insufficient supply of these trees.

The wood produces medicinal aerosols that will protect clothes from wool-eating moths. The wood can be used as a veneer lining on all manner of furniture for this purpose. The trees could be bioplanned as farm set-asides. There is a match of trees for every growing zone from zone 10 into zone 4 for this role.

The Canary Island juniper, *J. cedrus*, and the East African juniper, *J. procera*, are both trees that do exceedingly well in zones 9–10. These trees grow to 33 m (100 ft.). Their timber and wood are medicinal. The smaller trees of the red cedar, *J. virginiana*, the Colorado red cedar, *J. scopulorum*, and the alligator juniper, *J. deppeana*, could be part of a two-tier agriculture where these trees are introduced as hedgerows to cut up hyper-large fields. The crop could be twofold, grain and selective timber. These trees would conserve water in arid areas. Across the globe the drooping juniper, *J. recurva*, and its closely related cousin, the Afghan juniper, *J. squamata*, could be planted as two-tier agricultural trees also.

DESIGN

The juniper is the most flexible evergreen for garden design. The various cultivars that are in the horticultural market will serve every possible purpose in the garden. There are variants and forms that hug very closely to the ground. There are shrublike cultivars, some with elegant vaselike shapes and others that weep. There are many that can trail over a wall in a most attractive way and more that will imitate a lawn. There are tall columnar junipers that fit well into a small garden. These same species can be clipped into the wonderful puff-ball images of a Japanese garden. The tiny junipers can fill the evergreen spots in an alpine garden and even a scree or gravel garden should not be without its green delights. The juniper will grow in any soil that is well drained.

The taller junipers of the red cedar clan, *J. virginiana*, have a temperature cut-off of zones 3 to 4. The most subtle of these in its smooth upward shape is *J.v.* 'Pyramidiformis', which matures to 33 m (40 ft.). Then the Chinese juniper, *J. chinensis* 'Mountbatten', takes over to zone 2. This gray-green cultivar has proven itself over and over again in cold Canadian gardens to be a survivor. All of the cultivars of the red cedar, *J. virginiana,* work very well in a North American garden. These species are spectacular in a dry design. However, they cohost a blister rust, *Gymnosporangium juniperi-virginianse,* of apples, *Malus,* so they should not be planted where there are apple orchards. If apples are to be used with red cedar cultivars, then apples of mixed Russian ancestry will be less prone to the rust fungus. Many of the heirloom apples are in this category. They can be whitewashed for additional protection.

The ground huggers of note for formal landscapes and gardens are the Irish juniper, *J. communis* 'Hibernica', and for color, the steel-blue Montana creeping juniper, *J. horizontalis* 'Montana'. The famous shrub called the Pfitzer juniper, *J. chinensis* 'Pfitzeriana', is thought, by some horticulturalists, to be a wild form of juniper that came from the Ho Lan Shan mountains of Inner Mongolia. It certainly fits the bill, because this specimen juniper is an excellent windbreak for very cold temperatures and winds down into zone 2. In these temperatures the evergreen tops brown slightly in winter. The trailing forms for a gravel, rock, or scree garden are better found within the *J. horizontalis* cultivars, one of which is the *J.h.* 'Coast of Maine'. This gray-green juniper turns into a wonderful purple-tinted foliage form for the winter months, as does the *J.h.* 'Wiltonii'. These cultivars are called blue rug junipers because of their summer color of a startling blue. These cultivars change color, too, for the winter months into a deep red-purple, especially on calcium-rich soil.

Any of the Savin junipers could be designed into 'A Canine Loggia'. The resins and aerosol compounds from the Savin group are very beneficial to dogs. They should be allowed to roll freely on these species to improve their skin condition, which in turn produces a better coat. The top three are *J. sabina* 'Cupressifolia', a female clone with a high ratio of biochemicals, the more commonly available Blue Danube juniper, *J.s.* 'Blue Danube', and the Hicks juniper, *J.s.* 'Hicksii'. This tough triad of junipers will not bear a grudge against any dog if used for an odd roll. They will return the favor by producing a new flush of attractive shoots in the spring.

The hidden beauty in the cones of the decid- uous conifer called the tamarack, *Larix lari- cina*

ARBORETUM BOREALIS

Larix

TAMARACK, LARCH

Pinaceae

THE GLOBAL GARDEN

A tree commonly called the tamarack or larch is the water baby of the conifers. This tree sits like a rebel in a pool of water and snubs its nose at its eight remaining rustic sisters worldwide. The tamarack has learned how to survive in bogs and fens, muskegs and marshy woods. These are places where no self-respecting conifer should be found. But then the tamarack, otherwise known as the *Larix laricina,* is no ordinary tree. It is a leaf-shedding deciduous conifer, and that strange annual event came to pass some 250 million years ago. The reason, known to us today, is a twist of survival called biodiversity.

In the global garden there are only nine deciduous species of larch. Some are to be found in the Boreal circumpolar north, while others line the sides of mountains in countries such as Nepal and Tibet. However, the tamarack, *Larix laricina,* is the most numerous tree and makes a wide brushstroke across the Boreal of North America, making a turn up into the Yukon and Alaska. In the south of the Boreal the tamarack is the fairest of trees, with a plumb-line trunk aiming for the stars, but in and around the waters of James Bay the tree winds into the bent-basket shape of a shrub. A long time ago the mountains of British Columbia reinvented the genes of the tamarack and reprocessed the tree into an incredible species. These trees are the western larch, *L. occidentalis,* still holding the same geotropic plumb-line of 61 m (200 ft.), the height of a twenty-story building, and an age that registers a span of almost one millennium apiece. There also exists an alpine sport, the Lyal larch, *L. lyallii,* named for the good doctor who first cast eyes on this western tree at the 610 m (2,000 ft.) alpine tree line.

The Boreal of northeastern Russia and western Siberia carries the flag of the tamarack genes into the blinding cold of their north, expressed as the Siberian larch, *L. sibirica.* The Siberian tree fights the freezing air with a fur coat protection. To survive, the tree forms a fur of trichomal hairs together with a continuous wet suit of rubbery, glabrous skin. Finally, there is the grand stand of the great Kuril larch, *L. gmelinii* var. *japonica,* seen on the hills of China but vital to the sea life of Sakhalin and the rosary of the Kuril isles. Here in this birthing room of great mammals, the larch shoots grow heavy with the down of trichomal hairs as they turn the familiar reddish brown to face the first spring frosts for protection. This larch pays no lip service to the savage air swinging off the fetch of these raging seas.

All over the global garden the larch is the keeper of mountain streams, of rills and water sheds. These are the northern sources of fresh water for the planet that meet to form lakes and feed rivers. Water is of itself a simple substance, but it is etched into every biological system for life. It is as important to the trees of the forest as it is to man and his family. Without water there would be no reproduction for these two kingpins for life, because for both, water enables the act of fertilization to happen by providing the watercourse for the final flagellated swim of the male to the female ovum. It is upon this simple substance, water, that life is built.

The nature of water has been changing in recent years. This has been observed by the elders of the First Nations of North America and most probably by the elders of other ancient cultures worldwide. They all say that "the water has lost its light." These cultures are dependent on the oral tradition of word-to-word passage from generation to generation among their own

Tamarack, Larch

kin. They believe that water is changing, and they are right in this. Their observation of light in water is due to the molecular physics of the purity of water, just hydrogen and oxygen, together with the intramolecular forces that bind each molecule of water together. When the white light of sunshine passes through a body of clean pure water, it travels through the water itself in refraction. These rays of refraction serve to amplify the original light passing through the water, and so it appears to a viewer as if the water has a life of its own. And therefore there is more light in water by amplification. All pure water does this. The effect is increased if water is being aerated by an increase in the partial pressure of oxygen. This superpure water with the sharp edges of light was taken for granted all over the planet once upon a time. And, indeed, what the aboriginal elders have observed is the degradation of habitat caused by the felling of forests. This has changed the character of fresh water. It appears to have lost its light.

MEDICINE

The inner bark is the major medicine of the tamarack. It can be used either fresh or dried. Its medicine is more potent in the early winter months just after the first killing frosts. The outer reddish bark is medicinal, as is the outer bark of roots, in addition to the fresh roots.

Tamarack manufactures and produces a great number of highly labile terpenoids and other aerosol compounds. Some of these are in the extracted oil of tamarack, while others are not. Many of these aerosols have a high antiseptic and antibiotic activity. These are difficult to characterize into individual biologically active biochemicals, and it is even more difficult to trace their synergenicity together. Bornyl acetate is produced by the tamarack in large amounts as a carrier compound to aid the efficacy of the antiseptic aerosols. In addition, many tannin polymeric compounds are produced by the tamarack. Galactic acid, the brain sugar, is copiously produced, as are various forms of abscisic acid.

Little is known medicinally about the remaining larch species worldwide. The inner bark of northern tamarack is used to treat both burns and frostbite. This is probably because of the content of galactic acid in its polymer form, acting as a healing mucilage. Either fresh or dried cambial bark is chopped into as fine a powder as possible. This is applied to the burn or frostbitten area. Application is first thing in the morning, and the poultice is worn all day. It is removed at night and is reapplied as necessary in the following days with nightly release for skin aeration.

A hot-water distillation of the inner bark is also used to wash cuts and open wounds. This is also used as an eye- and earwash. And a decoction of fresh root is used as an aid to skin healing.

Tamarack is often used as an aboriginal heavy-gun medicine for the more serious diseases of venereal origin, for circulatory and heart disorders, and for treatment of rheumatism and arthritis. Severe colds and coughs are also included. In many cases a root decoction is added to the roots of the fern, *Osmunda claytonia*, the club moss, *Lycopodium lucidulum*, an *Equisetum*, a *Rubus*, a *Lonicera*, and a *Diervilla*. This decoction is concentrated down to a few milliliters. While it is being used as a medicine, a strict taboo is placed on alcohol and sweet food in any and all forms.

To increase its expectorant capacity tamarack root is added to others of the pine family such as the white pine, *Pinus strobus*, or the northern hemlock, *Tsuga canadensis*, when available. Here in a more concentrated mode the medicine functions in a manner similar to the air around these living trees, by acting as a cleaning aid to the working lung.

In the past, tamarack has been used as a high-protein food source in times of scarcity. The inner bark during the hungry winter months carries a considerable food value due to storage of sugars, fats, and proteins because of the deciduous nature of the tree. The inner bark is harvested. It is air-dried before it is pounded into flour that is very high in both fat and water-soluble vitamin content. As a food this bark flour has a high calorific and electrolytic content and is health-giving to the major organs of the body.

ECOFUNCTION

The honeydew produced by the tamarack and other larches considerably adds to the health of the northern migratory songbirds who use the circumpolar Boreal forest. This flush of sugar is amplified into insect protein, which in turn is a food of high value for the nurturing of the young. Healthy young birds will in turn then be capable of successful migrations southward in that long search for those special places on the face of the continent that seem to control their lives.

The larches are different from all other conifers because they have learned to manufacture a series of important hormones. These hormones regulate the canopy of the entire tree in addition to modifying root growth underground. The hormones are a group of abscisic acids, and they exist in various different spacial forms within the biochemical body of the tree. The molecule can take a swing to the right or to the left and act like a traffic warden switching on or off the abscission of leaf fall, shutting off photosynthesis during the long months of winter.

As the days shorten into the fall in the circumpolar north, the light levels of the sun begin to decrease. For the larch and the tamarack this is a time of change. The full body of the tree senses this loss of light and begins to produce a hormone called abscisin 1. Then a further mix is produced in abscisic acid. These hormones trigger the flow of chloroplastic green out of the leaves into the storage areas of the tree. The leaves have left only the xanthin color of orange to yellow. These are the colors of fall seen in the changing tree. And soon this, too, goes into the great food matrix of the medullary rays of the tree's cortex. Then a sharper frost or a drop in temperature alerts a binge of abscisic acid, which flows to its target. This abscisic acid gathers at the base of the leaves just at a point where the leaf detaches itself from the stem. This attaching string of tissue is called the petiole. A series of concentric ringlike cracks occur at the bottom of the petiole at the point of attachment of each leaf and its branch. Soon the leaf becomes a dead weight. As the wind rises, this dead weight begins to shift. The shift imposes a perfect solution in the physics of a circular crack, and the entire leaf floats away without any injury to the branches of the tree. This release is leaf fall, and it is the phenomenon of the deciduous tree. All that remains is the sealed leaf scar. It is easily seen on every tree. In fact each leaf scar is like a thumbprint. It is used in classification as a family trait for identification, one that never fails.

The manufacture of abscisic acid in the conifers of the larch family has a far-reaching effect of global importance. As fall reaches the circumpolar forests, the larches obey the laws of nature and lay down their leaves. They also switch off the carbon-to-carbon exchange of their root systems. When the green canopy has fallen, only the dormant skeleton of the tree remains. This tree is asleep. It does not sequester carbon dioxide out of the air, nor does it produce and spill oxygen into the atmosphere. This dormancy is felt in the rest of the global garden as a steep decline of atmospheric oxygen. There is an increase of oxygen again as the heat of the sun fires deciduous trees into action for spring. And once again more oxygen is produced to aerate the atmosphere to continue life. The switching on and off of the deciduous trees is like a lightbulb, with a respiration curve unique to the global garden. This curve has been changing this century. The plateau of spring oxygenation has been flattening out because of the relentless cutting of deciduous forests.

The deciduous nature of larches means that these trees aid in combatting global warming. They do this by the huge spring sequestration of carbon from atmospheric carbon dioxide into the mainly carbon skeleton of the leaves. This carbon is then deposited onto the soil under the trees as forest floor leaf mulch, to accumulate in depth. This becomes a carbon sink because its rate of decomposition is extremely slow in the Boreal. This sink is added to annually because of the deciduous nature of much of the conifer forest of the north. This life process drains enormous amounts, megatons, of carbon dioxide out of the atmosphere and sequesters it into a sink of carbon that is reprocessed by time, involving millions of years, into fuel or coal or diamonds.

Tamarack, Larch

The tamarack, *Larix
laricina*, and water-
sheds have always
been together.

BIOPLAN

The multitudes of tamarack, *Larix laricina*, that sit in muskegs
and sphagnum-laden swamps in the Boreal cannot be replaced.
The tamarack is the most numerous deciduous conifer of the
north. The tree has modified its internal biochemistry to survive
in a very specialized habitat of open, full solar exposure and wet
feet. The wood is resinous. It produces an oleoresin that is fully
waterproof. This makes all of the wood unique. In the past, the
pioneers made good use of the tamarack for their water systems,
pumps, and barns.

The new pioneers located water for their farms by witching. A
hazel, *Corylus*, rod was cut. It had to have a chicken bone shape.
The two handles were held by the witcher or water diviner, and
when a decent flow of underground water was located, the hazel
rod moved. If the presence of underground water was strong, the
hazel moved like lightning downward to indicate the water
source.

Then the pioneer went looking for tamarack. A tall, straight,
slim bole was taken. This wood was peeled of its bark and
worked into the form of a water pipe. This wooden pipe was laid
down into the freshly drilled well. The drilling was done using
either a dog or horse. The wooden pipe was connected with a
hand pump. The pioneer had the miracle of fresh water.

But another miracle happened a century or two later when the
well piping was being replaced. The tamarack wooden pipes
were hauled up out of these wells, and they were, for the most
part, seen to be in perfect condition, for tamarack can survive
almost indefinitely underwater. And the water was not contaminated because of the antiseptic oleoresin content of the wood. In
addition, the pioneers used tamarack rafters for almost all of the
cedar log barns they built. These, too, can be seen standing a
handful of centuries later.

The aboriginal peoples already knew about tamarack and
water. Their birch bark canoes were stitched together with the
resinous root fibers of tamarack. The fresh epidermis of the root
was peeled. This was retted into a fiber suitable to hold the birch
bark pieces together. The fiber was insoluble in water because of
its high oleoresin content. Then the peeled tamarack root, with
its right-angle bend, was cut to fit the inside of the canoe for rib
attachments. This method was copied by the large sailing ships
that plied the Atlantic. The tamarack roots were used for joining
the ribs of the hull to the deck timbers. The larger tamarack
roots were much sought after for this important task and were
an international trading item.

The tamarack and the larches of the Boreal act with their
remaining siblings to protect the fresh water of the landscape.
They are extremely important and indeed vital components of
watersheds. The remaining larches, the European larch, *L.
decidua*, the Chinese larch, *L. potaninii*, the Himalayan larch, *L.
griffithiana*, and of course the Dahurian or Kuril larch, *L.
gmelinii*, all protect mountainscapes. Their presence on slopes or
in mountainous terrain helps to provide a system of water flow
control that will become more important with the violent storms
predicted by climate change. These areas of forest worldwide

should be immediately placed under a no-cut order. The lessons of North Africa, Mexico, west coast America, Canada, many parts of Europe, and China all tell us in hindsight that steep slopes should never have been logged. Today we need water, not erosion.

The tamaracks of the circumpolar forest protect the ground-water of their own habitat. In the spring the trees sprout a fresh canopy. This canopy cools the surrounding soil by 2°C. This cooling acts as an air conditioner and reduces the surface temperature of the soil, which decreases evaporation. This cooling effect reduces the activity of the soil mycorrhiza, which in turn slows down the decomposition of surface leaf litter, which further acts to keep the soil cool. This decreases the outward flow of carbon dioxide from decomposition. The atmosphere benefits by a reduction of carbon flow, and the moisture levels of the soil are retained.

Industrial extraction of the European larch, *L. decidua*, produces a product called Venetian turpentine. This is a liquid resin that solidifies into a clear glass. It is used in the science of bright-field microscopy. It may also have a future as a light-transmitting material in the solar industry. The other larches, including the tamarack, have not been investigated for their resins in this field.

Beekeepers could plant tamarack and the native larches as bee pasture. These trees produce, in some years, copious amounts of honeydew that is high in galactose sugars. This honey is thick and is very dark in color. The comb honey produced from this source is uncommonly good. It is an old method of increasing honey pasture, and the conifer honey produced was called *tannehonig*, which was much used as a health honey in the European market in the past.

The wood of the tamarack is tough and resinous. The grain is light brown with a dappling of white sapwood. This wood makes an exquisite interior finish to a house and is equally beautiful when used in cabinetwork. The wood is used as an interior finish in boats also. Outside of this use, tamarack timbers are used for posts and sills because the wood is durable in contact with the soil. And of course tamarack still is being mined as a forest product for pulp.

DESIGN

The native tamarack, *Larix laricina,* is a most beautiful medium-sized tree. It has a fairylike quality in the garden, changing from soft green in summer into the palest of yellows in the fall. The tree itself, being bare in the winter, has a form that is always attractive. The cascades of tiny cones, which have turned upright to empty themselves of seeds, are dark brown and fill with ice crystals or snow. The snow line on the fine branches spells out a delicate tracery of the tree that is redefined by anything growing in the background.

The tamarack grows well on high ground, too. It is a good frost-hardy addition to any garden, city or country. The tamarack was taken over to an English garden in 1739. There it crossed with the native larch, *L. decidua,* to produce a weeping form of seedling that has since then been much admired. The cross is *Larix laricina × pendula.* It is also called the weeping larch. This magnificent tree grows to be quite large. It has long branches with pendulous branchlets and terminal shoots. These are glabrous and an attractive soft pink when young. As the fall rolls about, they change into a becoming deep purple color.

The European larch, *L. decidua,* and its bulkier cousin, the Japanese larch, *L. kaempferi,* are grown in gardens all over the world. These are both striking trees with a strange lacy effect in the foliage, which seems to fall down to and sweep the ground. The only larch that has been hybridized for common garden cultivars is the Japanese species, *L. kaempferi.* This has an attractive sport with blue foliage called *L.k.* 'Blue Haze'. There is also a smaller blue bun-shaped bush ideal for small city gardens or rockeries, *L.k.* 'Wolterdingen'. And probably the most magnificent larch as a lawn specimen is the *L.k.* 'Pendula' with long elegant weeping branches, a surefire showstopper.

Black spruce, *Picea mariana,* with its crop of stunning purple cones

62

Picea

SPRUCE

Pinaceae

THE GLOBAL GARDEN

The spruce is the workhorse of the global forest. This conifer sits as a jewel in the crown of the world, in a vast northern woodland called the Boreal. The Boreal forest includes Alaska and Canada west to east, taking the north of Europe into a northern smear of the continent of Asia, stretching all across Russia, ending with the Bering Sea. The northern islands of Japan are also part of the Boreal, as is one of the most precious places on earth, the *Ostrov*, or island of Sakhalin. This spruce-rich island protects the krill-rich thundering waters of the *Proliv*, or strait of Tatarskiy, the birthplace of the great Pacific whales.

There are other dibs and dabs of spruce forest seen in higher elevations. In all of these cases the various species labor to make much out of nothing.

The spruce has about forty-five relations that thread themselves throughout the evergreen global system of forests. Of these there are two dominant species that launder the cold atmospheric world of the north. These are the white and black species. They in turn have sired many genetic offspring.

The white spruce has a necklace of common names. These are Canadian spruce, Black Hills spruce, pasture spruce, skunk and cat spruce. This tree is honored by the Chipewyan, who call the white spruce the "Big Brother," thus claiming the affection of family comfort. The Latin and universally accepted scientific name for the white spruce is *Picea glauca.*

The other dominant spruce is the black spruce. This invaluable conifer also has many common names: eastern spruce, bog spruce, swamp spruce, double spruce, and short-leaf black spruce. The aboriginal world of the Cree named their most important tree the black spruce, *Mistikōpikī,* or spruce gum. The Latin name is *P. mariana.* There are many species of spruce across the world. In addition to the two dominant species, there is the commonly known Norway spruce, *P. abies,* of central and northern Europe, widely planted everywhere for the Christmas tree market. There is the tough Serbian spruce, *P. omorika,* used in traffic-troubled urban areas and the hilarious tiger-tail spruce, *P. polita,* of Japan, whose prickly character makes the tiger's tail a far easier thing to clasp than the living tree itself. There is also the Sakhalin spruce, *P. glehnii,* with its chocolate-brown bark and host of protective chemicals inside its blue-green needles that are more important to whale offspring than we will ever know.

The spruce is an evergreen conifer. Evergreen conifers roll their leaves into a needle shape. The needle may be long or short, but they are always stiff. In the spruce they arise from strange peglike bases that are always spirally arranged and are permanent fixtures on the tree's branches. These bases, in turn, make the branches rough to the touch. The trunk of a spruce is tall and straight leading into a conical canopy of green branches. Its overall growth is by means of a cambial skin of tubby cells found inside the rough trunk. These are high in food value for the tree. Underground, the root system does what it can to survive with a short taproot system and an extensive network of adventitious roots that go shopping on a daily basis for the nutrition for the mother tree.

The increase in height is maintained by the seasonal formation of a terminal meristem whose delicate presence is protected by the guardianship of tough watertight bracts. This meristem is a hormonally rich plug of cells that dictate the pattern of size and shape, which is unique for every spruce tree, much like the thumbprint of the human hand.

The provenance or history of the spruce is an ancient one in the global garden. Conifers like the spruce arose after the great splash of life of the Devonian period, which gave rise to the jungles of monstrous ferns and mosses. These were far taller, bigger, and greater in number than any tree we see around us today. A living remnant of this era is the strange fern called the Tasmanian tree fern, *Dicksonia antarctica,* which still survives in the peaty swamps of a few tropical gardens. These living fossils tell of a time of humidity, heat, and oxygen-poor air. They are the prehistoric ancestors to the spruce who rose to greet another dawn.

The spruce have seen it all before. They have gone through climate changes in the past. They have seen hot and cold climates. They have endured high levels of carbon dioxide and low levels of oxygen in the atmosphere. But following the Carboniferous era, when the oxygen levels rang a loud note of life into the world, rising to 35 percent oxygen in the atmosphere, the conifers really got down to work. They began to evolve again, this time into the flowering forest trees and shrubs we recognize around us today in our own downtown gardens. All of the trees of the forests are evolutionary kings sitting on top of the entire plant kingdom. They bestow oxygen; their oxygen is a gift of life.

This is the balance within nature. The tree produces the oxygen that the human family breathes. That is why the trees of the forest are considered to be the lungs of the world. And within that forest the evergreen conifers are like an added insurance system with pledges to keep the planet's atmosphere running despite the tough times in their own economy of growth. This, in a nutshell, is the importance of the Boreal belt that floods the northern world with oxygen, which sweeps downward as the north wind, as bracingly fresh air, into the southern regions of the global garden.

MEDICINE

The medicines of the spruce are to be found in the leaves, buds, female cones, trunks, and newly formed branch tips. The various resins of the wood and bark, both fresh and dried, are also medicinal.

All species of spruce are members or the pine or *Pinaceae* family. This is a tribe of trees long famous for the pharmaceuticals and important biochemicals they manufacture on a seasonal basis. As the spruce grows into maturity the chemical manufacturing increases within the tree itself. The slower the tree grows and the more marginal the circumstances of its growth, the greater the medicines the tree will hold. This holds especially true for all spruce of the Boreal belt, where growing conditions from south to north mint an increasing treasury of medicine.

Each spruce has around twenty-one to twenty-five medicinal biochemicals, all of which are important. The numbers of these biochemicals will not vary for each species of spruce, but the ratio and quantity of them will, depending on the particular site on which the spruce tree is growing. In addition to this there will be a variation in the numbers of isomeric forms of each of these biochemicals. These, too, are produced in direct relationship to the tree's overall environment.

There is also an important change in biochemistry depending on where the spruce is growing with respect to the globe itself. The North American continent produces a dextro-rotation in many biochemicals, whereas the Asian continent including Europe produces a levo-rotation of the same biochemical. For all practical purposes this left-to-right movement of a chemical may be of little importance except to the human mammal, which can often only process the dextro form as a medicine. The other version may cause potential trouble if taken in excess.

Probably the most interesting medicine of the spruce tribe is the resinous gum that is produced in great quantity by specialized resin canals within the wood of the tree itself. This gum is also found on all growing tips of the branches as a protective sheath from cold weather and as a shield against insect infestation and browsing. This gum, when used as a chewing gum, is a cardiotonic and helps the oxygenation of the blood as it circulates, especially during exercise. The gum when used as a tisane is antihypertensive and will reduce blood pressure. It is antianginal and will help with the circulation within the beating heart as it oxygenates its own musculature and enabling it to act as a more efficient pump. The resin gum of spruce is also anti-arrhythmic and will help the individual myocardial cells with their electronic message system of communication, one area with the other in the different geographic regions of the heart. In the past and in the present time, spruce gum has been used as an aboriginal endurance medicine for running or other tasks that physically stress the physiology of the body.

All spruce conifers are monoecious; that is, they produce two types of cones on one parent tree. The male cones do not last long and are identified by their powdery drooping appearance. These cones are not used in medicine. The female cones, when in their juvenile state of being fully closed, green, and still encased with resin, are boiled in well water. The strained water is used as a medicinal mouthwash. This is used for sore throats, toothaches, or mouth infections. It is also used as a gargle to clear phlegm from an infected throat.

Spruce gum was and still is used as a handyman salve for the skin. The gum was mixed with animal fat. The favorite and most delicate source for the aboriginal populations was bear fat, but any fat including petroleum jelly will solubilize the biochemicals out of the gum. This antiseptic cream was used for skin infections, various rashes, and chronic skin sores. It was used as a disinfectant covering for burns and most commonly used for dry skin, especially in winter, when the natural oils of the epidermis are at an ebb.

Antiseptic and nontoxic baby powder was obtained from mature spruce wood that was allowed to rot. The dry wood was ground down into a fine powder. This was sprinkled on tender skin to protect it from the sore rashes of teething.

ECOFUNCTION

The spruces of the Boreal forest help to cure air sickness on a global scale. This sickness is new. It is a form of particulate pollution of the atmosphere. The air itself contains a moving body of tiny particles of 2.5 microns or less in diameter. These are smaller than pollen grains. The air itself is contaminated with these microscopic particles. They arise from industrial pollution, urban traffic sources, warfare, and from some natural sources like active volcanos. These particles travel more actively in warm trade winds and less actively in colder areas.

These minute particles, or PMs, become more deadly as they travel. Depending on the electrical charges of their floating surfaces, many organic molecules can stick to those surfaces like glue. Some of these include heavy metals, pesticides, and other mixes of monomers. These are spilled out of exhaust from traffic and many industrial processes. These particles as they become more toxic are turned into molecules that become death traps for the human body and all other air-breathing creatures.

Essentially, polluted air is inhaled for its life-giving oxygen. Both the oxygen and the particulate pollution travel in the same pathway. They go from the mouth to the lungs to the tissues of the body that need oxygenation. This oxygen transfer happens in tiny blood vessels called arterioles. These tiny arteries kiss the tissue with life, but for the oxygen to pass through to the tissues, the arterioles must relax. Relaxation comes with nitric oxide that is around to complete this job of relaxation. The presence of particulate pollution hampers or stops the job of nitric oxide. Therefore a constant state of inflammation is set up.

This inflamation affects all kinds of things in the body. It can reduce the live birth weights of babies and aggravate asthma. If the pollution carries a load of heavy metals it can lead to the thickening of the walls of the carotid artery and to strokes and heart attacks. In live recordings of blood pressure taken in response to real-life concentrations of airborne particles of traffic exhaust of 2.5 microns or less, the diastolic pressure of the heart has been shown to increase by 9 percent. This form of pollution compromises the web of all life on the planet.

Trees, forests, and especially spruce trees of the Boreal comb the air of some of this particulate pollution. The living spruce needles function like the teeth of a comb. The needles themselves

Boreal forest floor lichen, the reindeer lichen, *Cladina arbuscula*, is tinder dry.

are covered with a waxy, resinous cuticle whose complex biochemistry supplies a chemical bonding attractant for the airborne pollution, thus directly removing the particulate pollution from the air. This marvelous ability puts the forests of the world, and most especially the Boreal, front and center for the direct protection and cleansing of the air we breathe.

The conifers like the spruce of the Boreal forest have modified their growth to their habitat. They have developed an extraordinary adaption of the thermodynamic reaction that captures the energy of sunlight. This reaction is the photosynthetic reaction. In reality it should be temperature-dependent like all other organic reactions of the biochemical world. It is not. This is because conifers like the spruce have become smart over the millennia. The leaves conserve water by their shape and by means of a thick cuticle of wax covering them. Each breathing pore is also modified to reduce water loss. The pores are called stomata, and there are millions of them. These conserve water for the tree. In addition to being water smart, the spruce can photosynthesize in the shade at lower temperatures of $23^{o}C$ down to $9^{o}C$, which is lower than most other trees. In addition, the spruce hangs onto its leaves by being evergreen, thus reducing its seasonal need for food. Food with nutrition comes by way of a shallow root system that creeps just under the surface of the soil with just the right amount of exposed root to press the root mycorrhizae a little further to produce. This forces the maximum nutrients and food exchange out of a minimal system. This kind of adaption multiplies efficiency in a poor habitat like the Boreal north.

The soil structure in which the Boreal forest has to survive is less than good. Poor drainage and the presence of permafrost together with high levels of moisture saturation makes for growing conditions that are difficult at best. In many parts of the Boreal belt the presence of igneous and metamorphic parent rock gives rise to the acidic character of the soil itself. The strongly acidic nature of the soils inhibits the action of soil mycorrhiza associated with decomposition. This lack of rapid natural decomposition leaves an accumulation of litter on the Boreal forest floor. This leaf litter as it accumulates dries out on the surface. As a dry surface mulch, it, too, conserves available moisture underneath.

In these dry conditions another set of organisms move in, smelling nitrogen. These are extremeophiles, tinder dry for most of their life cycle. They are the lichens. The most visible of the Boreal lichens are the old man's beard species of the *Usnaceae* family. They inhabit the surface of the forest floor. They occupy the dry lower limbs of the conifers and move up the trunk in their hunger to fix nitrogen. They give back so much more to the trees than they take. They, too, are tinder dry, the thallus structure open for firelike kindling.

While the dry balls of lichen live on the leaf litter and the leaf litter sits on top of the moisture below, all is well for the Boreal. But it is thought that global warming will make dry places even drier and wet areas more moist. As the arctic air warms up, so will the leaf litter and the lichens, too. Fire, the normal mode of spasmodic nutrient recycling in the cooler Boreal, changes into something deadly. The leaf litter of conifers, especially species like spruce, still holds explosive chemicals like camphor, which is commonly used in pyrotechnics. With the advent of climate change and concurrent drying and warming, the burn of the Boreal will be vast and fast. This will cause significant atmospheric change by flooding massive amounts of additional carbon dioxide into the air we breathe, making it increasingly toxic to mammals like us.

Another important ecofunction of the Boreal forest for the entire planet is that this vast crown of woodland acts as a health barrier. It does so in a surprising way. Trees like the spruce are loaded with an arsenal of biochemicals, many of which are in the form of aerosols. The spruces also manufacture an array of dispersing agents, in such quantities that the pine family, of which spruce is a member, is extensively used in the perfume, disinfectant, and deodorant industry. It should come as no surprise that this is what conifers do. Trees like the spruce disinfect the air of the atmosphere with antiseptics. Medicinal biochemicals such as antibiotics, too, are released. These sterilize the air, targeting mycobacteria in it.

The species of the spruce are particularly committed to this behavior, in a most clever way. The mature leaves of the spruce produce an alkaloid that is an amine called ethanolamine. An increase in airflow around the mature needles of the canopy releases this amine to travel around the lichen mantle of the tree. This amine has a strange function: it triggers the manufacture of antibiotics of the lichen species. The lichens manufacture them as part of their marginal household chores. Antibiotics are then liberated from lichens. These are added as hitchhikers to the carrying aerosols of the spruce, and both carry a message of health into the airways of the Boreal.

Then the spruce does something else. It manufactures a sticking compound called beta-phellandrene, which is a chemical glue for the skin. And so the antibiotics find their targets on exposed skin to fight infection. It is all carried in the form of fragrance. This biochemical is also found in the essential oil of spruce. It is why the resin gum is medicinal and it is also why the species of the Boreal are good for global health, even when the circumpolar spruce forests are so far away.

The Boreal forest facilitates the north-south migrations of birds worldwide. Trees like the spruce in spring produce pollen-laden staminate cones. The pollen is high in the essential amino acids for protein building for insect life. In addition the spruce is a source of honeydew, a by-product of insect damage. This helps feed soluble sugars into the high demands of the insect world. Resin gums from the spruce are used for insect housing as medicinal wallpapering and to keep the home of the insect both warm and dry. Insects accelerate life up the food chain of the forest to birds who fly north to feed and build up their resources for egg laying.

Avian migration for successful reproduction is a life force of the planet. It has been going on for so long that the memory of it is embedded deeply in the genetic code of the bird. Birds are beneficial. They are the first line of defense through insect predation in the far-flung fields into which the members of the flock will fly. This is true in Europe, North America, and Asia, where birds will do the task they have always done, that of grooming trees and other plants of pathogenic insects.

The conifer forests of mixed spruce are loaded with an arsenal of beneficial biochemicals.

The Boreal belt as a northern carpet reduces global warming by affecting radiation. This vast northern carpet of conifers is deep green. The green of the conifer needles is so dark in some species that it is almost black. The color of the leaf is a functioning living tissue that is called mesophyll. In the conifer this mesophyll is dense and extremely concentrated. It is within this mesophyll that photon capture from radiation takes place. The density of the mesophyll is tied in with its efficiency: the darker the needle, the greater the capture.

This greenbelt reduces the greenhouse effect of global warming. The incoming solar sunshine is made up of short-wave energy because it has come from such a hot source. This energy hits the greenbelt of the Boreal and is reduced. The resulting wavelengths become longer and more useful to the trees. In addition, the dense mesophyll of the conifer needles resonates this energy within each leaf as a heat reserve before it gets ferried off chemically into the photosynthetic reactions taking place in the matrix of the mesophyll. This reaction in turn produces enormous amounts of atmospheric oxygen while it reduces the carbon dioxide load.

67

Spruce

The steady branches of the black spruce, *Picea mariana*, are an open invitation to the world of the lichen.

ARBORETUM BOREALIS

Because the Boreal belt is so efficient in its use of radiation, there is less long-wave radiation, called infrared radiation, remaining to bounce between the carbon dioxide shield of the atmosphere and the earth itself. The effect is like a tennis ball that has received a puncture and will not bounce. And so the air temperature of the atmosphere is not increased. This radiation-absorbing skin has a reducing effect on global warming for the entire planet, north and south.

BIOPLAN

The black spruce, *P. mariana*, the white spruce, *P. glauca*, and the red spruce, *P. rubens*, species are marshaled together to produce pulp for paper. The demand for paper is enormous in every literate society. The introduction of new telecommunications technology has not decreased the need, rather the opposite. As the East meets the West in the modern consumer society, the need for pulp will rise again. The success of the spruce family has been the means of its own demise. The little tracheids that both plumb and feed the growing tree are strong and thick-walled. The spring tracheids are thinner in wall structure than the fall tracheids, but their fate remains the same. These are the fibers of which paper is made with all of its grades and quality and ease of coloring.

Another blessing of the spruce family that has also been turned into a curse is the bole of the tree, which does not show much of a difference in the cross-section. In other words, the pith area is not too different from the outside wood, and this is good for the pulp and paper manufacturers. At the present moment there is very little serious research and development on a pulp substitute for the spruce family and even less agricultural advice given to the farming community to plant small parcels or set-asides of spruce with the specific intention of selling the lumber for pulp as part of the agricultural industry. A small grant to plant these trees and care for them would go a long way to help the mixed family farm globally and to induce thinking about sustainability of the land.

The long fibers of hemp of the *Cannabaceae* family or the yucca of the *Agavaceae* family or boiled bark of the *Tiliaceae* family, when added to recycled spruce pulp for the recycled paper trade, would make a better paper product. This could be recycled indefinitely and make paper itself more sustainable in a growing market. Added to this would be a benefit to a farm with poor or little soil. Such soils would produce a tougher pulp fiber of hemp, yucca, or basswood with more schlerenchymic cellulose than on better, richer, ground. This in turn would contribute to a cleaner planet and reduce global warming by protecting the standing spruce forests of the Boreal north.

All spruce that are cared for and grown with a soil surface mulch will outpace nonmulched trees. This can also be done on poor stony soils. The rate of growth is increased up to 100 percent. This can be a useful maintenance tool for a farm set-aside, where many forms of mulchlike spoiled hay, straw, or manure mixtures become a problem. They can be used as mulch to surface dress a small spruce plantation to great reward.

Spruce produces a highly valued lumber for the construction industry for framing, sheathing, roofing, scaffolding, and for use in subfloors. Spruce is also widely used in the manufacture of construction plywood. And the wood is used for the food industry mainly because it is aseptic and nontoxic. Slow-grown, air-dried spruce is used for the sounding boards of many musical instruments.

Industrially, spruce pitch is used as a nontoxic sealant. The oil of spruce is extracted from the by-products of the lumber and paper industry to create a product called *absolute* for the perfume industry. This absolute is used as an aromatic and antiseptic in soaps, cosmetics, and many common household cleaning products.

As the damaging effects of weather changes proceed to produce chaos in many parts of the world's housing stock, there will be an increased demand for all spruce. Set-asides in the bioplans of the farming community would provide some planning for the future both here and abroad.

A beer somewhat similar to Irish Guinness can be made from

spruce. The young tips, each with a meristematic bud, are added as a spray to the fermentation process. The result is a dark beer not quite so black as the licorice-fermented Guinness drink. But it is an equal in its health-giving abilities, containing micro amounts of all that is beneficial in the spruce resin and pitch.

Probably because of its ancient provenance, spruce manufactures its own insecticide, which is particularly active against cockroaches. The compound is 1,8-cineole. And, curiously enough, this compound is also found far away from the cold Boreal belt. It is found in high concentrations in the eucalyptus trees of Australia and Tasmania, where maybe there were high populations of cockroaches once upon a time.

A pitch product called Burgundy pitch was extracted from the European spruce, P. abies. This deep red-colored pitch was used to treat lumbago, rheumatism, and chronic bronchitis. It can still be seen today in another form, in old hand-fashioned violins. This pitch was used as a surface sealant. Its tight elastic movement with sound gave the Stradivarius of Cremona, Italy, its wonderful mellow tone, causing shrill auction prices that have soared in pitch throughout the auction houses of the globe.

DESIGN

The spruces form an extremely ornamental group of evergreens for the garden. They are drought-tolerant with highly desirable xerophytic characteristics that fit into any xeriscape, a method of garden design compatible with global warming, when water will certainly become more scarce.

The spruces contain a wide range of shapes and sizes. These go from medium-tall trees to compact dwarfed cultivars. The foliage varies also, going from the deepest green through silver shades to a positive blue color. Spruces thrive in a variety of soils, which can be both poor and shallow.

The white spruce, P. glauca, and its cultivars will tolerate chalky soils better than most, as will its European cousin, the common spruce, P. abies, which has a wide range of dwarf cultivars, the most popular one being the marvelous German sport, P. abies 'Nidiformis', or bird's nest spruce.

An attractive spruce that likes an acid soil is the North American east coast native, the black spruce, P. mariana. This tree, with its gray-brown bark and matching young shoots, makes an elegant statement in any garden with its host of purple-colored young cones. This tree does better in moister soil and air, as does its Chinese cousin, P. likiangensis. This medium-sized variable tree is at its peak in spring, when the green needles are contrasted by the brilliant blood-red color of the new season's crop of female cones.

Probably the most admired spruce is the Brewers Weeping spruce that hails from Oregon. P. breweriana is a breathtaking sight as a specimen in a large garden. This small to medium-sized and broadly conical tree needs both space for its weeping habit and for its many admirers.

Rock gardens and small city and urban gardens have many minute cultivars to choose from. Probably the most flexible and interesting miniature is the hedgehog spruce, P. glauca 'Echiniformis', which presents its prickly character in combination with another and more important one, that of a means of escape for ground-feeding songbirds from the salivary glands of the neighbor's cat!

A bioplanned set-aside of white spruce, P. glauca, is probably the best species for a rural setting, a farm, or poor land on which an easy cash return is required in the future. The trees should be well spaced one from another. The space should be mulched to give maximum growth potential to this lumber- and pulp-producing tree. This species has the added advantage of being ignored by deer herds and avoided by porcupines. These trees could be selectively harvested for either pulp or timber as a cash crop.

The jack pine, *Pinus banksiana,* is some of the last wildwood left on the planet. It is to be found in the Boreal.

ARBORETUM BOREALIS

Pinus

PINE

Pinaceae

THE GLOBAL GARDEN

Pinus banksiana is the princess of the Boreal north. This pine is a transcontinental species that stopped about 500 km (300 miles) short of the Pacific. The British Columbia Rockies were in the way. But another related pine, the lodgepole, *P. contorta*, picked up the ticket and finished the journey to the seaweed of the shore. There it sits in full view of the tides, a twisted variant named the beach pine, *P. contorta* var. *contorta*, holding the sandy shore away from the onslaught of the sea. The doppelganger, *P. sylvestris*, the Scotch pine, holds a similar vast landscape, that of the Taiga forested region of Russia from the Bering Sea to the Baltic Shield.

The pine called *P. banksiana* was not previously known to the settlers and pioneers who stepped foot on this continent. They saw a magnificent tree and were sure it required a deep rich soil on which to grow. Their land grants were in their hands. But they were wrong. The pine and its soil they had inherited forever turned out to be the bare bones of rock. So they cursed the pine. Then they named it the jack pine in one another's company. The name stuck.

They were sufficiently impressed by this species to make sure that it was passed on to the parkland of their landlords' estates back home in England. The date was 1735. The empire was rising like yeast. The dough of success was fueled by the never-ending planks and timbers from the exotic forests of North America, which were eternal and would never ever run dry.

So for the *P. banksiana* the first common name of jack pine was followed by a lot more, banksian pine, scrub pine, black pine, and for the French, gray pine. The jack and lodgepole pines, *P. contorta*, are so closely related that they will intermingle and hybridize to give progeny whose seeds are in turn, viable. However, the aboriginal peoples like the Déné of the Athapaska and the Algonquian, Anishabe, and Algonquin who live with and use this pine in their everyday world of survival in the Boreal forest call it *kohe*. It is a word of echo, an old word from the Déné language of the Slave nation of the Athapaskan north.

In the global forest, *P. banksiana* is important. It is one of the approximately one hundred pines that have mantled the earth. This pine has a taste for fire. It was acquired in the past, so much so, that the lessons learned are now part of the plant's genome of survival. The banksian twin cones will continue to survive on the tree side by side for up to a quarter of a century or more. These cones remain closed, the bracts withdrawn and tight. The viable seeds are held in the sealed humidity of each vault, safely, until excessive heat or fire comes into the season's weather pattern. Then the cones will open and release. The black-eyed seeds will take to the air propelled by a single slimline propellor. This tailored propellor is capable of lift into warming air. The rising currents activate the finely balanced tailspin for dispersal, to journey considerable distances well beyond the dangers of fire. Once landed, the little seed will wait out the winter to germinate in the spring. Then it produces a unique, handlike cotyledon set of leaves.

As an adult, *P. banksiana* can easily be identified by its stunted leaves or needles. These needles are among the shortest of the pine family. The needles are paired to form an angle, and each needle has a sharp tip. This configuration makes the tree prickly to handle. The fertile cones, too, are in pairs, each female cone on either side of the branch. They do not attain a right angle in growth because they are banana-shaped. But they do straighten out on the dispersal of their seeds. The mature canopy

Pollenation time for
the jack pine, *Pinus
banksiana*

of the tree is oval in shape. The aging trunk webs itself into an attractive formation of furrows and islands of darker plates.

P. banksiana with its love of fire has an eye for change. The tree colonizes, whenever it can, into open burned-over areas. Like many colonizers these pines are short-lived, making way for other forests. They will grow for 150 years and more. In good sites with deep sandy soil, they will live longer and attain heights up to 28.5 m (90 ft.). But in other places of less hospitable habitat nearer the edges of the tree line of the Boreal, they do what they can to survive, becoming stunted and dwarfed.

P. banksiana represents some of the last wildwood or virgin forest left on the planet. These areas of incredible growth are still to be found within the Boreal belt, where a tree is not just a tree—each is a metropolis for other species. But when taken as a whole, the living landscape creates a tableau of an extraordinary set of species with molecular diversity, one that certainly cannot be re-created.

Because of its Boreal boundary, *P. banksiana* has married the world of lichens. This marriage is based on nitrogen. It is a give-and-take affair where the cyanoalgae component of the lichen mops up aerial nitrogen and turns it over to the fungi-manufacturing part within the lichen home. The lichen in turn acts as a marriage partner to the tree, colonizing it from, and including, the ground upward into the branches and even the twigs. This affair of symbiosis frees an organic form of nitrogen to the tree in return for the tree sugars to fuel the fungi. This happens around the roots, too, but the feeding pattern here is a little different. At the roots other cyanoalgae species of the mycorrhiza operate in a more naked form to attend to the deliveries of the tree.

The saturation levels of the ground and the permafrost of the Boreal make the life of the soil mycorrhiza more difficult and push the limits of what the lichens can and must do.

Of the families of lichens that love the Boreal world, perhaps the *Usnaceae* family is the greatest. This family of lichens is called the old man's beard. This is because the lichen body trails down off the branches of trees and looks for all the world like an aged and grizzly beard on a wooden face. The various grays of the *Usnea* and *Bryoria* species that are common to Canada are replaced by the similarly functioning beards of *Ramalina* species in the remainder of the global circumpolar Boreal region. This huge biomass of suspended lichen manufactures usnic and salazinic acids, among many others. These acids with their accompanying fixing agents add a considerable bolus of antibacterials into the atmosphere. These chemicals released from lichens act as an air purifier, maintaining clean air by their antiseptic action. And the chemicals themselves are capable of dispersal like an oil film on water.

Over the millennia, all peoples both east and west in the global garden have considered the pine to be a medicinal tree. It was used in health resorts and for specialized hospitals. The pines, with humid warm air, possess anesthetic properties. The inhaled fresh air exerts a dilatory action on the bronchi. The inner tissue of the lungs gets coated with a miracle of esters. These go into the bloodstream to fight respiratory diseases, and the pines of the Boreal do it best of all.

MEDICINE

The medicine of *P. banksiana* is to be found in the fresh needles, in the pine gum, and in the cambial living tissue of the inner bark of the mature tree.

The needles of *P. banksiana* can be used to treat frostbite. The species that are growing either on the poorest or most northern habitat within the Boreal woodland contain the highest ratio of chemicals for this kind of nerve and tissue regeneration. These needles contain two diterpene resinous acids. These are concentrated by drying the pine needles down. These are then crushed into a brown powder. This powder, when applied to the frostbitten area, will help with tissue regeneration.

There are a few plant medicines that hold active ingredients for nerve regeneration. *P. banksiana* is one, and the common snowdrop, *Galanthus nivalis*, is another. The snowdrop is more effective as a medicine when it is taken from the snow line in mountainous areas, where it is higher in the regenerating biochemical called nivalin. Outside of the possibilities of stem cell research, these two naturally occurring nerve regenerators are important to science.

In addition, the two diterpene resinous acids of the needles of *P. banksiana* have the ability to promote lactic acid fermentation, that is, the fermentation of milk and yogurt products into their high-protein cheeselike form. There is also a butyric acid fermentation. This would in turn lead to a search for novel yeast strains of *Saccharomyces*, whose adaptability and importance to industry is legendary. These would be found in the Boreal.

The pitch from pine has been used all over North America as a treatment for colds, flus, and infections. The jack pine, *P. banksiana*, carries the greatest mixture of essential oils of them all. The gum itself was chewed as an on-the-spot cold medication and preventative.

The soft inner bark of *P. banksiana* was further softened by soaking it in warm water. This bark was worked until it was as soft as a bandage. This was then applied to a fresh wound. If the wound was deep and the healing had to happen from the base upward, the jack pine bandage was continually reapplied as a poultice until the wound began to close and heal over.

ECOFUNCTION

The jack pine, *P. banksiana*, and a wood warbler known as the Kirkland's warbler are bound forever in life. This large warbler with its yellow breast and wagging tail searches out the pine needles of the jack pine to line its sunken nest. Both male and female return north in the spring from the Bahamas, looking for a large stand of pine. This stand must be in the region of 12 hectares (30 acres) of young trees of less than 7 m (21 ft.) in height. Here this songbird will reproduce and carry its low-pitched soundings for life.

Another wood warbler called the Northern Parula warbler is even more fussy than the Kirkland species. This rufus-capped, tail-wagging songbird goes for the lichen called *Ramalina*. This bird on its return from Honduras in the spring uses the thallus structure of this lichen to line its nest in the ground. *Ramalina* is chosen over its cousin *Usnea*. Both lichens are pendant epiphytes on the branches of *P. banksiana*. But the *Ramalina* species of lichen has a propensity to grow on trees that are in turn feeding off silica-bearing rocks. This small switching of geology and the biochemicals produced both in tree and lichen is vital to the life of this songbird.

P. banksiana is a feeding tree for the insect life of the Boreal woodland. The tree as young as four years old will produce multiple male cones at the apex of its growth. Their load of pollen is in heavy demand as a source of food. This increases insect reproduction and has a cascading effect on all the wildlife around this species.

Across the North American continent jack pines, *P. banksiana*, and the lodgepole pines, *P. contorta*, produce aerosol and chemical carrier compounds that dispense antibacterial or antibiotic medicines into the atmosphere. The anatomy of each tree is designed as an open dispensing agent to free these chemi-

The strange twin
cones of the jack pine,
Pinus banksiana

cals into the air. They come from the mature bark of the trunk, from the resinous branches, and even from the leaves. When this effusion is taken together with the trees' epiphytic lichen load, the total of the combined biomass effect is considerable. The gross effect is to cleanse the air, purifying it using the arsenal of aerosols and carrying biochemicals loaded with natural antibiotic agents as air fumigants. This protective load acts as a disease shield for the world of the north, moving into the south with the northwest winds of early spring and summer.

BIOPLAN

The same quality tracheid fibers that are diminishing the spruce forests all over the world are also to be found in *P. banksiana* and *P. contorta*. This is an unfortunate state of affairs for these trees, because the word *pulp* again arises and the insatiable demand for it. On poor shallow soils *P. banksiana* grows in a snarling twisted way. The fragrant wood is laden with resinous knots. These trees are ignored. The virgin species growing on deep sandy soils that soar into the heavens are chosen, trees with their boughs trailing their biomass load of lichens almost to the ground. These are the trees that are going down for pulp, cut by newly designed scissor machines each worth six million dollars. The operators are taught how to operate these machines in virtual classrooms. They are mostly young aboriginal men looking to a future that will vanish with each cut and every virgin tree felled, a betrayal of their sustainable way of life.

As a bioplanned species the *P. banksiana* has an enormous growing range on the American continent. Beginning south of the Arctic Circle, the range is some 1,600 km (1,000 mi.) south and 4,000 km (2,500 mi.) east to west. Within that arena *P. banksiana* can be grown by the farming community as a set-aside for a multipurpose system. On shallow rocky soils, *P. banksiana* in its stunted form will grow the most aromatic wood,

resinous branches, and internal connecting knots with the greatest amount of pitch. The leaves, too, will have a thicker waxen cuticle. These could be harvested for pine oil extraction for the perfume and cosmetic industry. Even their bark will serve this purpose. A burn-over and reseeding would tweak the Kirkland warbler habitat and keep it in good shape.

The larger timbers and planks of *P. banksiana* are relatively waterproof because of their high levels of resin. This wood has been used in the past by aboriginal peoples for boat building. It could also be used for a similar purpose in the future, planned and planted as cash crop set-asides.

The jack pine can be grown on good land for both the pulp and wood market. This set-aside could act as a cash flow system for agriculture if the need so arose. Grown under these conditions, the trees could be a source of hydro poles, railway ties, and the usual wood products for the building construction trade. Because the wood is in itself aromatic and nontoxic, it can be used for packaging and then reused.

The aboriginal peoples of the Algonquian nation and the Athapaskan group could use the virgin wilderness to their advantage. These areas could be used in a restricted tourist trade in a manner similar to the tourist management of Bhutan, where tourists pay heavily for the privilege of occupation. The tourists are strictly limited in numbers, thereby maintaining the sustainability of the area while maintaining the cultural diversity and governance of the north. The northern aboriginal peoples need a better cash flow for their own livelihood in a place where they are nature's stewards.

In addition, as the southern portion of the continent loses its rural character and goes toward the urban mode of living, the wilderness of the Boreal north becomes more important. The native knowledge and hands-on living skills of the aboriginal world could fill a niche of spirituality and culture that is increasingly missing in the urban society of the south. This stability of the north could be a two-way street that could benefit both north and south immeasurably.

DESIGN

P. banksiana, as a single specimen tree, is airy and beautiful. It seldom loses its needles and is thus tidy in its habit. This tree is open in its growth, being somewhat flat topped at maturity. This moves the growing plane of the tree into an interesting visual horizon in the garden. Both open and closed cones on the tree add visual interest. This is not a tree for chalky soils, as the leaves turn to a yellow color in the winter months with this kind of habitat.

P. banksiana and its west coast cousin *P. contorta* var. *contorta* are both excellent trees for dry areas and as such are ideal trees for garden design during the water restrictions of climate change. They are part of a drought-tolerant or xerophytic landscape with little or no effect on surrounding shrubs and flowering species. These are ideal trees for areas of severe water restriction, where they will cool the ground and help maintain the water table.

Apart from one dwarfed pendulous cultivar that is not generally available, *P. banksiana, P. contorta* var. *contorta,* and *P. contorta* var. *latifolia* are grown as their native form. However, the branches can be trained into espalier patterns quite readily for a Japanese garden style or a site near a water garden for reflection.

An aspen, *Populus tremuloides*, canopy filtering the sun

ARBORETUM BOREALIS

Populus

ASPEN, POPLAR

Saliaceae

THE GLOBAL GARDEN

A weird little group of trees is known all over the global garden because of the movements their leaves make in the slightest breeze. These trees are commonly called the aspens or poplars.

This genus is represented by a group of about thirty species. They are all tucked into the famous willow family, *Saliaceae*. Aspens are astonishing species because they are living windmills in the world of trees. Action is in their blood. It can be traced to the physics of their design to catch the flow of the softest puff of air. The design of the leaf with its long stem or petiole is like the pendulum of a grandfather clock. But the motion for movement is wind generated.

The aspen, *Populus,* is found throughout the circumpolar forest. In Canada the glistening stands of trembling aspen, *Populus tremuloides,* are called *wasī-mītos* by the Cree, which means the "bright poplar" because of the reflected light these trees amplify from one to another in a stand in the fall, so much so that they appear to be self-illuminating. A quieter tree is found growing in the Siberian Boreal called the laurel poplar, *P. laurifolia.*

Outside of the arctic zones the aspen, a long time ago, was witness to Christ in the Middle East. There, in the torrid heat, the aspen turned into the Euphrates poplar, *P. euphratica,* a tree the locals called *Gharab-Palk-Saf-Saf.* This little treasure is even found in Kenya. In Europe, northeastern Asia, and northern Africa, another aspen called the trembling aspen, *P. tremula,* is seen to quiver and shake. This tree marched along the ancient footpaths of Ireland, nodding its head so much that the druids noticed and called this tree *crann creathach,* which implies that the aspen quivers so much that it sets up a vibration like a shudder of emotion.

The Japanese, Chinese, and Koreans are witness to the same shudder in their Asian variant, *P. sieboldii,* which is also called *P. tremula* var. *villosa.* The Japanese aspen has a different rock and roll, coming from its bearded bottom and shaven face—in each leaf, of course. Finally, the balsam poplar, *Populus balsamifera,* is treated separately because of its importance in the global garden.

The sexual life of an aspen takes the high road into adventure. Not satisfied with the usual male-to-female rapport, these trees have veered into something new. Not unlike the clonal patterning manipulated by stem cells, aspens produce clonal groupings. Some are of an enterprising size. They consist of many thousands of identical trees covering many square miles or hectares in size. These mono-forests of ancient vegetative growth, where each cell of each tissue of trunk, root, and leaf is identical to another in this extraordinary mass march of existence, may well represent some of the oldest living organisms on earth. It is thought that these gangs of like-minded trees began life 1.6 million years ago just after the ice sheet melted during the Pleistocene era.

In addition to the quirky clonal culture, these trees transgress even further and practice the gay life in homosexuality, something, too, that would appear to have been around since Dikika man, the latest find in fossil man, *Australopithecus afarensis.* For some, items in the sexual life of biological systems will not go away, whatever the mores of the day. Maybe, because the aspens are a very common group of trees worldwide, not much thought is given to the importance of this species, that is, outside of the pulp industry in papermaking. However, the aspen is an important medicinal tree in the pharmacopeia of the aboriginal peoples of the planet.

The slender aspen stands with a gray-white bark whose

smooth powdery veil falls down to a rougher, darker footing. The base of the tree is nearly black, and the bark in this area balloons to a thickness of 5 cm (2 in.). It is rough with braids of warty growth. The powdery white bark as it climbs up into the air wears slashes of darkening scars to match its feet. It is only in the southern Rocky Mountains of the North American west that the aspen finds a true welcoming home and rises to the bar with a majesty of 30.5 m (100 ft.) before diverting into its egg-shaped canopy. Here in this airy scene, the diameter matches a more robust growth and the tree's girth swells to nearly 1 m (3 ft.). Outside of the mountains the aspen resumes a more temperate size and also matches age, seventy years, with that of its foe, the logger.

The leaves of the aspen tell their own story, as they have throughout the ages. Few cultures have been blind to their shaking or deaf to their teasing noise. Every aspen leaf is attached by a petiole. Compared to other trees, this petiole is very long, far longer than the leaf. The petiole length for each tree never varies, for it is set by the genetics of the tree. The petiole is attached to the tough leaf surface. The petiole shape is pinched in the middle. This pinching of the three-dimensional form helps to hold the sheath of the leaf forward in a position to cup the faintest breeze. Under the petiole's point of attachment to the leaf are two counterweights of semicircular collenchyma tissue that govern the volume of air trapped, causing movement like the pendulum in a grandfather clock. The trichomal whiskers residing at the edges of the leaf pick up airflow like the hairs of a listening human ear and carry forth sound with movement.

MEDICINE

The medicine of the aspen is to be found in the bark, in the mature leaves, in female flowers, and in juvenile seeds. It is also found in the inner bark and white cambial tissue. Because the tree as a unit is an efficient electrolyte collector, the ashes of the semiburned wood were once used in North America as a substitute for table salt, sodium chloride. It is found concentrated in the wood ash.

The aspen, like all of the willow family, *Saliaceae*, manufactures salicylates. These salicylates represent a large family of chemical compounds all of which are related to one another because they carry a similar aromatic ring structure that can move throughout the body with great facility. These compounds are biochemicals that are all related to aspirin. Aspirin is the most widely used drug in North America, with over 50 million tablets consumed on a daily basis.

Each and every species of poplar, for example the trembling aspen, *P. tremuloides*, carries a differing ratio of salicylate biochemicals. This ratio varies with the growing season over the continental divide of North America and from continent to continent. The isomers produced by the poplars of China will not be the same as those of North America. The trees of the Boreal circumpolar forest will produce the greatest and most potent load of salicylate medicine because of the stress induced by the harshness of the climate. It is in the arctic north that the best of the best will be found.

The aspen contains the potent glucoside salicin, which can sail anywhere in the body with the aid of its water-soluble sidekick, glucose. It is the potent painkilling action of salicin that is the ubiquitous analgesic. In the alcohol form, which also occurs, this painkiller turns into a local anesthetic. The aldehyde, a yet more volatile form given off from the surface stoma of the leaf, functions as an aerosol, as does its cousin, the acetyl form. Both are so volatile that they are used in the perfume industry. They both dissipate very rapidly into the surrounding chemistry of the local atmosphere.

In the body the more volatile forms all travel easily in the fluids of circulation and become potent analgesics or painkillers. Their action is anti-inflammatory or inflammation-reducing. In this they easily crank back an elevated temperature, dropping it down to normal ranges and reducing the ravages of fever. This is antipyretic and is one of the most valuable medicinal aspects of the aspirin family, especially for the young and vulnerable.

Then the salicin changes to another face. It drops its sugar molecule and picks up an aldehyde grouping. This turns the chemical around into something very familiar, a kitchen creature commonly used in flavoring. The compound becomes vanillin. In the life of the tree as in a cake, vanillin is a flavoring agent. The taste of vanillin is as attractive to a moose as it is to a child. It excites the demon of temptation in the palate. Naturally, the industrial use of vanillin is extremely varied, from confections into the dairy industry. It is even added to liqueurs. In the business of flavoring, there are no bounds.

The aspen contains medicinally active flavanoids and an interesting array of active and flexible tannins. In addition, the active alkaloid tremulacin is in the mature leaves. The cambium may contain a form of papain, a folded proteolytic enzyme that is used in the treatment of contact lenses to prolong wearing time.

The aboriginal uses of the aspen as a medicinal tree are long and varied. The mature leaves were used as an on-the-spot treatment to relieve the pain of wasp and bee stings. This was also used to reduce the inflammations of mosquito bites. The fresh aspen leaves were initially chewed to soften them and then applied as a poultice to the skin.

The bark periderm, or outer skin, was scraped and a powder was obtained. This dust was used to treat deep wounds and put on surface cuts to stop bleeding. The periderm is high in tannins which are strongly hemostatic. Outside of the Boreal, aboriginal peoples of the south made use of this function also; a fresh root of trembling aspen, *P. tremuloides*, was added to another of the balsam poplar, *P. balsamifera*. Both were steeped in cold water. This drink was used to prevent premature birth with problem pregnancies.

Trembling aspen was also used as a treatment for serious heart conditions. The southern treatment was ritualistic and involved the inner bark from the bur oak, *Quercus macrocarpa*, the red oak, *Q. rubra*, and the trembling aspen, *P. tremuloides*. This was added to equal amounts of the root, bud, and blossom of the balsam poplar, *P. balsamifera*. The last ingredient was a root fragment of the Seneca snakeroot, *Polygala senaga*. The medicine was prepared ceremonially, with the first two ingredients being dried and the next two, the poplars, being powdered by hand. One-half liter (1 pint) of water was the volume used. To the north, south, east, and west a pinch of the first four ingredients was placed on the water's surface. Then a small pinch of powdered polygala root was placed on the tree extract. This was allowed to steep. The patient was handed a birchbark dose receptacle. The patient was prepared to drink only one swallow. An hour later this was repeated with the same careful attention to the volume. This latter was a second and carefully timed "heart-shock" treatment.

This process was modified for the Boreal world, where a length of aspen bark was cut the same size as the human heart. This was taken from the south side of a mature tree. It was removed at the height of the beating heart. The bark was chewed to solubilize the medicine, and the juice was drunk. This simpler form was also an effective heart medicine.

Aspen bark was used to treat diabetes, toothache, stomachache, coughing, cancers, diarrhea, and venereal diseases. Lastly the young seeds of aspen were collected and chewed. They were swallowed as a means to induce abortion.

ECOFUNCTION

The aspen is a host tree for the insect world. The pollen of the aspen arrives early and is eagerly looked for by native bees and flying beneficial insects as a food source for their brood. Because of the high mineral content of the tree's sap, especially magnesium, zinc, copper, and molybdenum, the pollen is a source of co-enzyme factors for the insect's metabolism. The sap is also useful as a food source for birds, insects, and butterflies. In the springtime the flying squirrels of the north seek out the fresh elongating buds just as they are at the right stage of sugar conversion. The buds are eaten and the epidermis bud scar oozes sap. This sap becomes an open food market to the aerial brigade of local and migratory species.

Aspens, *Populus tremuloides*, enjoying the mixed company of conifers in a northern alpine habitat

Aspen, Poplar

The many butterfly species that use the aspen as a host species are the dusty wings, the admiral, the viceroys, and the ubiquitous tortoiseshells, among others. These pollinators are all in search of the same thing. The electrolyte balance of minerals in the aspens supplies the butterflies with their requirements for specific metals. Each butterfly is unique in its demands, depending on its wing color requirements. The metal is ingested and then passed on to the correct biochemical pathway for metal chelation. When the metal is set, it becomes a color, either visible in the normal spectrum or a color invisible to humans within the ultraviolet range. In any case, wing color is vital to butterflies for their courtship displays, which lead to sexual success and reproduction.

The butterflies of the Boreal woodlands and elsewhere look to the aspens for something else. The northern tree species have the highest content of bitter-tasting salicylate complexes. These biochemicals offer a toxin trick protection from predation. This protection aids the safe passage of migration and protects the lives of butterflies, allowing them to reach sexual maturity. Thus the beneficence of the aspens to the insect world forms a backdrop to successful cross-fertilization of other plant species associated with the aspen, *Populus,* species across the globe.

The aspen itself is an opportunist. The tree is one of the first to jump into the vacuum left by forest fires, by abandoned fields or open places. The aspen moves either as a seedling or underground as a clonal root sprout. But the tree is armed for warfare. The chemical load that the tree carries is an antiseptic one. It clears the way by its life and by its death for the regeneration of the conifer forest that shortly follows in its path. The conifer comes in armed with its own mycorrhiza in a steady state for the long haul in time.

Each and every aspen leaf of every aspen tree's canopy operates individually in the manufacture of salicylates. They are found in increasing concentration in the mature leaves. And each leaf acts as an applicator. The mass of trichomal hairs that line the sawlike serrated edges of each leaf is infused with salicylates. As the air moves, the leaves flutter. Their motion is multidimensional and releases the salicylate aerosols into the atmosphere. Sometimes they emerge as esters, other times as alcohols and aldehydes. The overall effect is that the salicylate aerosols act as an antiseptic for the atmosphere. This air rises out of the circumpolar Boreal forest to scrub the atmosphere of the planet.

The aspen is a medicinal tree also for animal life. As the leaves mature they produce vanillin, which is a taste enhancer. This happens at a time when mammals have been feeding. Some unfortunates have ingested the eggs of intestinal worms and parasites. These infestations seriously inhibit the health of the mammal population. So the smell of the aerosol ester of vanillin comes by their nose and the aspen leaves are eaten. The leaves are strongly anthelmintic and the worms are excreted by all manner of life including deer, coyotes, and wolves.

The aspen itself has been a source of food for a very long time. The inner bark and the cambial tissue can be peeled off into strips and eaten. These have the familiar sweet flavor of honeydew melon. The sap of the aspen can be collected in the spring as it rises. It can be made into a sweet syrup in the same fashion as maple syrup. Domesticated animals, especially goats, seem to have a fondness for aspen, as do sheep and horses. On the wild side the aspen is browse for deer, elk, moose, and reindeer. Beaver populations fell the trees to store the bark for winter. Rabbits and mice eat the bark, foliage, and buds. Songbirds use the cotton from seeds for nesting material. The larger overwintering birds such as grouse and quail, further south, feed all winter long on the available buds.

Ever since the Chinese invented paper making in AD 105 using bark mixed with flax, *Linum usitatissimum,* and pounding both together with stones until they had a mushy paste that was gently hammered into sheets and dried in the sun, the aspen has been on deck for pulp. Whereas the Chinese used their own trembling aspen, *Populus tremula,* all other species of aspen, especially those of the circumpolar forest woodland, are now up for grabs. It would seem sensible to leave the circumpolar Boreal intact as a major life force of the planet and look to other pastures. For instance, cloning matched to hybrid vigor of aspen

ARBORETUM BOREALIS

could be just one of many potential answers within the tribe of the *Populus* species for commercial exploitations.

BIOPLAN

The sex life of the aspen has produced a beneficial side-effect. The trees, *P. tremuloides* and *P. tremula*, on both continents have produced children that are triploids. That is, the burden of each cellular nucleus is greater with DNA than in other normal aspens. In the case of the triploids the cells carry three sets of chromosomes. But the case does not rest with this alone. The aspens with this triplex number are extremely disease resistant and grow exceedingly rapidly into fiber-rich trees. As luck would have it, the American and European species in all likelihood would have an affinity for one another genetically, and out of this transatlantic love match a superior tree filled with first-generation hybrid vigor could be bred. The trees from this match could be superior in every way for pulp and paper production.

The first generation of the transatlantic match could be vegetatively cloned and out of this mixture could arise disease-resistance, resistance to the stress of global warming, ability to withstand climate change, rapid growth, and hybrid vigor expressed with increased quality of fiber strength and length for paper. The fiber, because it could be larger in a more rapidly growing medium of a hybrid aspen, would also be easier to deal with industrially. The manufacture of quality paper products would require less digestion in the retting process.

Set-asides of species or hybrid aspen forests around cities as greenbelts would benefit industry as it would increase the health of the population by scrubbing the air with aerosols. The trees too would act as living filters for both particulate pollution and the reduction of smog. Greenbelt trees would also reduce the ability of smog to become airborne and travel further afield into other sectors of the globe. The greenbelt trees by their physiological act of mass transpiration increase the humidity index of the immediate atmosphere and add the essential water vapor to the particulate pollution, increasing the density and mass and reducing its ability to remain airborne as smog.

The aspen offers another jewel in the crown of civilization. This is its windmill function, which is seen in the movement of each and every leaf. There is an increased global demand for energy that is green and safe. This need is rising in the wealthy West but it is also of vital importance to poorer countries that are trying to get on their own feet and be self-sufficient. The nature of aspen leaves, particularly those of *P. tremuloides* and its kin, and their ability to move with so little airflow offer answers that are new to science. Nature has shown the way, from flight to Velcro to antibiotics. The nature of the aspen leaf indicates movement and movement is important for energy generation.

The exact scale model of the leaf can be enlarged into another working model for movement. The energy of movement can be trapped by molecular film and transformed into work. Work initiates electron flow, which can be captured and used. The jump from idea to initiation is never a big one, and the aspen tree can indicate one way forward for a greener planet.

DESIGN

The aspens, especially *Populus tremuloides*, present a great beauty in their bark in any garden. The white sheen of the outside periderm acts as a contrast to all other native trees. The aspen can be planted alone as a single specimen, or it can be planted in a spaced grouping of either three or five. As a group the fall color of butter yellow is magnified and acts as a curtain of color in the garden against which all other fall colors will shine.

Populus tremuloides produced an exquisite little tree as a sport in a garden in France about a hundred years ago. It is a weeping form called 'Parasol de St. Julien'. This tree has the added benefit of being a female. It does not produce pollen and therefore is a nonallergenic tree for a sneezing household. This tree is a perfect addition to a school or a kindergarten's bioplan.

Aspen, Poplar

A spring fingerprint of
the female balsam
poplar, *Populus bal-
samifera*

82

Populus balsamifera

BALSAM POPLAR

Saliaceae

THE GLOBAL GARDEN

There are mythical trees and there are mystical forests. Some are lost. Others make themselves known out of the mist of time, like the Wollemi pine, *Wollemi nobilis*. This chance population was found in 1994 deep in Australia's Blue Mountains. This fishbone pine bears the fossil features of 200 million years. Then there is the *Metasequoia glyptostroboides*, straight out of the backdrop of China, with its fir face ready to stare down anything in Jurassic Park. The circumpolar Boreal forest has its share of fossils, too. One that is alive and well is so extraordinary that it should be treated alone. The name is *Populus balsamifera*.

The balsam poplar, *P. balsamifera*, is affectionately known as the balm of Gilead by most people of the North American continent. The Cree peoples of the Boreal call it the "ugly tree." But then, the Boreal balsam will not win any beauty contest. It is still around because it is smart. The balsam poplar has the same kind of cunning as its cousins, the fabled tree of biblical times, *Balsamodendron opobalsamum*.

Long ago in the ancient lands of Palestine there was a tree of miraculous properties. This tree was called the tree of Mecca, *Balsamodendron opobalsamum*. This tree produced a fragrant oleoresin. This resin was used in Egypt and it found its way as far as the Indian subcontinent. It was commonly used throughout the entire Middle East. The resin was processed into an oint-ment that became the basis of a strong import-export business that lasted until the early eighteenth century. Now the tree is believed to be extinct, although some botanists hope that some populations of it may still survive in the rural areas east of the Jordan River.

The early Semites named the tree. And so it carried the Hebrew name of *Báshám*, meaning an oily or resinous substance from plant origin. The balsam poplar, *Populus balsamifera*, of the Boreal also produces a similar kind of resin as an oleoresin. The Boreal balsam is a medicinal tree of the aboriginal native pharmacopeia. The ugly tree of the Cree has extraordinary medicines.

For the Semites the oleoresin product from *Balsamodendron opobalsamum* was so successful in its medicinal properties and was able to cure so many diseases that the tree was dubbed the tree of Mecca. As its fame grew, the remarkable ointment was thought to have supernatural powers. The cures seemed to transcend the mixed views of the local religions. These boiled over under the heat of religious strife with the result that the trees were cut down, all of them, for the unusual theater of war.

As the tree of Mecca came down, the balm of Gilead rose into prominence for the new pioneers of North America. The ugly tree or the balm of Gilead, *Populus balsamifera*, was indeed a tree that carried a fragrant oleoresin that did remarkable things, a godsend to the poor mixed farmer whose team of horses made a beeline for the balm of Gilead to chew the threat of colic away. The farmer would nod his head to the greater knowledge of nature and pass the news on to his neighbor.

The balm of Gilead is found growing across the entire continent of North America from Labrador in the east to the western slopes of Alaska. This tree lives in a reverse cycle in the Boreal north; its habitat of choice is just under the Arctic Circle in a climate that is far too severe for most other hardwoods. It attains its finest form in the northwest of Canada and can clock in at heights of 30 m (100 ft.) with trunks to match of 2 m (7 ft.) in diameter. In this cold habitat the mature trunk is always straight and tall, growing a good 13 m (40 ft.) before it splits up into its canopy.

Balsam Poplar

The balm of Gilead trees very often form pure stands. These are found on the banks of streams and on the borders of fens, lakes, and swamps. They are quite often found growing with alder, birch, spruce, and willow. The balm of Gilead produces male and female trees. The males have thin staminate catkins that wither in the spring after fertilization. The females carry furry pistillate flowers that fall like snow after pollenation in April.

The vast numbers of the balm of Gilead trees of the Boreal release a huge tonnage of oleoresin aerosols and vapors into the northern atmosphere. This acts as a health shield for the rest of the world when this tonnage mixes in the vortex of rising heated air from the south. This medicinal function is important for the health of the entire global community when an air-sweep helps to maintain the quality of gases that go into respiring lungs in all mammals everywhere.

MEDICINE

The medicine of the balsam poplar, *P. balsamifera*, is to be found in the sap and in the mature leaves. It is also in the buds of late winter to early spring, before they begin to break and to elongate. In this case the brown leaf scales are also included in the medicinal grouping. The medicine is in the female pistillate flowers as they have elongated and are just at the point of showing a white cotton at their tip. The medicine is in the bark of the tree both as it ages and also while young. This bark is removed as a north, south, east, and west fraction, all with different medicinal concentrations, the southern portion being highest in more concentrated medicine effects. In addition, the trees of the northwestern Boreal forest are the strongest and best medicinal trees in the global garden for the production of balsam medicinal oleoresins.

From a scientific and biochemical point of view the medicine of the *P. balsamifera* is extraordinarily complex. It truly is, as the medicine men of the Boreal forest advise, "a very strong medicine."

The medicine of the balm of Gilead that is currently known is not altogether scientifically understood. The oleoresins of the early spring brown buds contain 60–75 percent dihydrochalcones, including other flavones and flavonols. There are coumaric acid and cinnamic acid together with all of their esters and terpenoid complexes, which amounts in total to 11–13 percent. There are bisabolol and arachidonic acid, which is the essential fatty acid involved in brain, liver, and glandular development. There are prostaglandin derivatives of arachidonic acid. There are phenolic glycosides and salicin with its benzoate form called salicin benzoate. There is also the licorice-tasting populin.

In addition the propolis of the balsam poplar contains flavonoids, pinocembrin, and galangin. There are pinobanksin and pinobanksin-3-acetate. There are the benzyl esters of paracoumaric acid and the esters of caffeic acid. This propolis is a health food with natural antibiotic and antifungal activity.

The complex biochemicals such as the prostaglandin grouping have just recently been discovered in conventional medicine. Their biochemical metabolism of movement into more refined chemical action in the body, such as the production of a strange chemical called prostacyclin, is only vaguely understood. Prostacyclin alone has an effect on the innermost lining of the arteries, keeping blood cells moving right along and preventing them from clotting. In other words, it inhibits platelet aggregation that can be the beginning of all manner of circulatory trouble for the human body. This effect is found, too, in the coronary artery, the real Achilles heel of modern man.

The oleoresin also contains two essential nutrients for human and other mammalian health. They are extraordinary biochemicals whose molecular structures are shaped like open bullets. These bullets are made of fatty acids that can change shape into the most essential ingredients for intelligence. These are called linolenic and linoleic fatty acids. These are the building blocks of the brain, and when they are represented by arachidonic acid, they change to become the matrix of the liver, the heart, the brain, and all the glandular organs. In addition they form the unique system of fat tissue newly discovered. These are deposits of brown fat that handle the ability of the body to withstand cold by shivering. The shiver index helps the body to metabolize fat to fuel its furnace to survive.

This is just the beginning of the description of the oleoresin of the Balsam poplar. There are flavones and aromatic compounds that can float in the air as esters. There are perfumes and painkillers called salacin that can float in the air too.

There are strange licorice complexes in populin that have direct actions on the control systems of the stomach and digestive tract. Most of them occur together as glycosides. All in all, it will take many laboratories and legions of scientists to tease out the medicinal targets of the oleoresin that the balm of Gilead produces.

The Cree and the Slave nations have tapped the balm of Gilead in the early spring. The sap produced by the tree was used to treat diabetes, as were the buds, which were boiled with the bark of the younger branches of the trembling aspen, *Populus tremuloides*.

A tisane of balm of Gilead was used to treat heart ailments. Late winter buds were used. Two whole buds were steeped in boiling water. The tea was strained and taken as it cooled.

A unique treatment of psoriasis or eczema involved making a hot-water bath on which the spring buds were floated. The steam extracted the oleoresin, which floated as an oily layer on top of the water. The bather got into the tub and proceeded to rub the warm oil on the surfaces of the skin of the affected areas. This oil was allowed to dry in place as the treatment. The buds were also used in the management of toothaches and soreness in the mouth.

A south-side bark decoction was used to treat stomachache. It was also used to help the effects of seizures.

Balm of Gilead bark was added to butternut, *Juglans cinerea*. A decoction was used as a laxative. This medicine was used increasingly southward on the North American continent.

ECOFUNCTION

The ugly tree is the primary medicinal tree of the Boreal north. As the sun moves up in the horizon to shine on the buds of the trees in the months of March, April, May, and June the buds warm up. The dark protective sheaths of bud scales that have closed rank about the living apical meristems begin to change. The buds have been held tightly closed by the solid resin. This begins to melt. The movement triggers the sheath to expand. This in turn causes the bud to swell, moving the auxin hormones. These growth hormones stage the elongation of the bud into a branch. The balsam resin protects the entire system from the cold and, more importantly, from the ice crystals that can form within the living cell structures. These crystals can act like a blender destroying everything and anything in its path including the powerhouse of the nucleus with its DNA coils. The resin also sheds rain, keeping everything dry.

As the sun touches the donkey-brown sheaths, these warm up first. The act of warming changes the physics of flow of the oleoresin. Heat moves the resin molecules, and the flow begins. The more volatile molecules take off from the resin flow and become airborne as esters and terpenoid flyers into the atmosphere. The molecules of oil have also been changed by special airborne carriers that the tree supplies. And the oil of the oleoresin changes into a water form as the oil becomes more soluble as an ester, ether, or terpenoid form. The tree spills a greater load of oleoresin into the atmosphere as the arctic temperature cranks up.

By the hundreds of millions, each and every ugly tree, *Populus balsamifera,* of the Boreal woodland does this, and the stretch of it is immense moving across the entire continent. The heated atmosphere becomes saturated with an oleoresin that is an expectorant and anti-inflammatory. It is antibacterial and antifungal. This fine massive aerosol prepares a healthy path for growth. Every living thing benefits in this harsh world of the north. Even the soft bellies of the south benefit as the spring air of the arctic billows across the continent.

As the oleoresin evaporates, the sheaths curl back. They have been held in place by the stiffness of the resin. As the sheath leaves abscise and fall off, the outer layer of the green bud is exposed. It can begin to photosynthesize in the sun, charging the cells. Soon the growing tip feels the heat. The meristem at the tip

Medicinal spring buds of the balsam poplar, *Populus balsamifera*

Balsam Poplar

A sexual body of the *Phallaceae*, or phallus mushroom family, on the balsam poplar, *Populus balsamifera*

is ready to move. The density of the top cells shifts. The hormones for growth are released. Each lower cell gets a lick of auxin, the growth hormone complex. Spring is on the way and the tree responds by growing.

The world that the ugly tree lives in is a sealed jam jar, called planet earth. In this jam jar there is a give-and-take of chemistry that either breaks down life or builds it up. The web of this life process affects every single living thing from the bacteriophage to the whale. And everything in between gets touched by this pattern of living. The balm of Gilead cuts a wide swath for the life of the entire planet as the oleoresin moves into the atmosphere and forms good air for breathing.

Biological activities for which the oleoresin-laden air is responsible include a stimulation of all smooth muscles of mammals including dilation of the small arteries. There is a deep bronchial dilation followed by a lowering of blood pressure. There is a slowing down of gastric secretions of the stomach wall lining and a change in fat metabolism, especially in fat breakdown. There is an inhibition of platelet or red blood cell aggregation in the circulatory system, and the blood is kept flowing. There is an induction of the labor of birthing and also menstruation. Even in the eye there is a change of ocular pressure. There also appear to be changes in autonomic neurotransmission, a reawakening of the brain.

The trees of the forest, of any forest, either tropical or Boreal, act like a herd of cattle. One animal alone will not do too much, but when individuals are taken together as a herd, the complex changes and there is a herd instinct that is shared. The trees are like this. They have a shared instinct, and the instinct of the ugly tree is for protection, much like its long-lost cousin of biblical times, the tree of Mecca, the tree of miracles.

When the human family emerged a few million years ago, it did so under the protection of the balm of Gilead and other trees like it. The creation process that caused the ascent of man did it also for the tree. This tree, the balm of Gilead, *P. balsamifera*, enjoys one of the strangest alliances of the plant kingdom. This alliance takes place with one of the most complex and advanced

species of fungi. The fungus in question is a basidiomycete called *Mutinus curtisii* or *Ithyphallus impudicus*, of the *Phallaceae* family. On any damage of the balm of Gilead, a sexual revolution takes place. The fungus that lives within the balm of Gilead changes its growth. The asexual hyphae become sexual. The newly wed hyphae grow to form a penis above ground. It is always produced either near or on the injured root. The penis is the correct scale and color of an engorged member. As it matures in a day or so, the penis produces an overwhelming stink. This attracts carrion flies that cross-fertilize the penis mushroom. Basidiospores are produced. They look around for another balm of Gilead to infect. This cycle of life has been going on for millions of years, and to date this union remains unquestioned.

BIOPLAN

A national bioplan for the Boreal woodlands is to leave the circumpolar forest intact. The *Populus balsamifera* or balm of Gilead changed its genetics a long time ago at the dawn of man, well before the last ice age, and it decided to call the near arctic home. Why this happened millennia ago probably had something to do with past climatic changes that spun the coin for a colder climate for the balm of Gilead trees. Indeed, as they go southward in their growing habits they change tactics. Areas of southern balm of Gilead are thought to be the last representatives of very old growth. This growth does something different, too. The tree itself makes a decision to reproduce. It does not do it in a normal fashion with an egg and sperm, which is known as sexual reproduction. Rather the tree buds vegetatively. The tree puts out underground runners like root suckers that are somatic clones. The genetic code of the clones is the same in every single cell of the reproducing tree. In this way large single stands of balm of Gilead trees are produced, each identical to the other. Such tactics for survival usually happen when a species is under stress, which it probably was a long time ago. Now such remedies are represented by stands of forest that have identical somatic or nonreproducing cells. These are found in the south of the continent like long-lost sheep. These stands can sometimes represent many hectares (acres) and are a fingerprint of past perils to remember.

The commercial product known as Canada balsam is obtained from *Populus balsamifera,* which is also known as *P. candicans.* The Canada balsam is used in the pharmaceutical industry in the treatment of upper respiratory tract infections, in particular for coughs, laryngitis, and bronchitis. It is also used in an ointment to relieve the pain of arthritis and for the healing of cuts and bruises. Canada balsam is also an excellent gargle for sore throats. The importance of Canada balsam could potentially turn the tree into a cash crop tree for the farming community. A certain percentage of the buds could be harvested for medicine.

In addition, the *P. balsamifera* can be grown by the farming community as a source of lumber for the toothpick and match industry. The wood is also antiseptic and makes an ideal tooth sponge for the removal of dental plaque. A tooth sponge reduces the incidence of periodontitis because the plaque colony is manually disturbed every day. This reduces the desirability of the tooth as a place for the bacteria colony to inhabit and thus the health of the gum line improves. This specialized wood could represent a cash crop on a sustainable farm.

Urban forests and planning for city green spaces should incorporate the balm of Gilead, *P. balsamifera,* as an atmospheric cleansing tree. It cleans the air of particular pollution of 2.5 microns or less. The billions of glandular hairs or trichoma on the surface of the leaves help to comb the air free of particles. The female trees should be planted as greenbelts near schools, nurseries, senior citizens' homes, and hospitals. The female tree is nonallergenic because it does not produce pollen, but it still has the ability to produce aerosols of oleoresins for the continuing health of the community and its air.

DESIGN

The balm of Gilead, *P. balsamifera,* should be part of a North American medicinal garden, where the fresh spring odor of the growing buds can be enjoyed. The tree in a windy site releases its fragrance into the air and adds its stunning scent to a fragrance border.

Balsam poplar has a sport called *P.b.* var. *michauxii* that has both hairy petioles and underveins of the entire canopy's leaves.

As a garden specimen the balsam poplar should be mixed with other trees where it can be loved for what it does and not for its visual treats. The balm of Gilead is an ideal background tree and mixes well with other deciduous or evergreen trees as a windbreak or shelter belt.

Balm of Gilead is a very useful tree for the horse world. The aerosols of this tree are beneficial to horses. They can be designed into the landscape outside of a horse barn, or they can be planted as a portion of the hedgerow system of fields in which horses, especially mares and their foals, graze.

Balsam Poplar

Boreal chokecherry,
Prunus virginiana, in
bloom

ARBORETUM BOREALIS

Prunus

CHERRY

Rosaceae

THE GLOBAL GARDEN

The cherry has offered a temptation to man and beast for a very long time in the global garden. It was a favorite of neolithic man. The midden heaps of these times expose enormous numbers of cherry stone leftovers mummified by time.

The cherry is a stone fruit. That means that there is a sweet flesh surrounding a hard-shelled seed. The number and variety of cherry trees in the global garden is astonishing. There are over four hundred species of cherries. These mostly sit in the Asian corner of the global garden, where the rich soils match their flowery temperament. Over one hundred have been dragged struggling into cultivation. A few still roam free. These are to be found in the southern portion of the Boreal circumpolar forest.

For the Western world the cherry rode into prominence in the eighteenth century when Pierre Belon introduced the East to the West. He brought with him in his hot little hands the cherry stones of a particularly beautiful cherry called the cherry laurel, *Prunus laurocerasus*, which hails from Asia Minor. This evergreen tree was quickly grabbed by the gentry as a first-class ornament. The English landlords passed it on to their Irish country cousins, and soon their Palladian mansions were awash with cherry laurel. They were planted in Dublin, Ireland, of course. The year was 1731.

One bright, sunny day the Irish country cousins decided to give a ball for their English landlords. So the army of cooks got down to business to produce the ocean of food. The cooks ran out of bitter almond flavoring, an acquired taste from the Moors. Bitter almonds were at the core of their confectionary. So an enlightened head cook decided to improvise. She went out into the estate parkland and collected the cherries from the cherry laurel trees. She washed them, then peeled the fruit off the stones and washed them again. She pounded the stones with her stone mortar and pestle to get the proper bitter almond flavor. She did. But she poisoned the guests. The landlords were not impressed at the number of dignitaries in expensive coffins being drawn by the plumed black horses over the streets covered with barley-corn straw.

The aboriginal men and women of the Boreal north did not repeat this fiasco in the preparations of their local cherries for pemmican. They went one stage further and smashed the cherry stones together with the red flesh in their wooden mortar and pestles. This smashing liberates the hydrogen cyanide as a gas from the fruit stones. Then the subsequent preparations of cooking, destroyed the other available cyanides. This made the fruit of their beloved chokecherry, *Prunus virginiana*, perfectly good to eat. The nation of the Cree emblazoned the recipe into the name, calling the chokecherry the "berry that is crushed." And they had never even heard one mumble of the Dublin city doldrums.

The cherry is in the genus called *Prunus*. This is a very old name coming from the Greek *proumné*, for plum tree. All the fruits of this genus are divided into a further five based on how the tree flowers and bears its fruit. Chokecherry, or *P. virginiana*, is in the group that produces strings of flowers each of which is followed by a line of small cherries. The strings are called racemes. In the majority of the circumpolar forest the cherry trees have these strings or racemes of fruit. An exception is the pin cherry, *P. pensylvanica*, whose small sour fruit is carried in clusters much like the common supermarket sort, the sweet cherry, *Prunus avium*. The pin cherry follows the chokecherry in the Boreal habitat. Both sweep across the continent beginning with the Atlantic Maritimes. Neither one finishes the trail to the Pacific coast nor enters into the landscape of Alaska.

The Siberian Boreal is visited by the presence of a low-slung ground cherry to escape the ice and snow. This smaller imitation of the chokecherry is called *P. fruticosa* or the ground cherry. This mop-headed tree bears a crushing load of tiny, sour bird cherries. It also occurs in the European Boreal forest.

A second tree brings up the rear guard for the more hostile areas of the alpine regions of Japan and Korea. This is the Japanese alpine cherry, *P. nipponica*. This bushy tree produces small black fruit. The Kuril Islands have a more unusual species, *P.n.* var. *kurilensis*. This is a small, bushy shrub as well, but the flowers are pink and the many fruits are purple-black in color.

A second *Prunus* species tracks the *P. nipponica*, the Japanese alpine cherry, for Japan, Korea, and the Kuril Islands. It is *P. sargentii* or *P. serrulata* var. *sachalinensis*. This late March or early April blooming cherry bears hordes of food for the songbirds in their migrations. This special tree is the first to color the Asian autumn with tints of crimson and orange, much like its chokecherry cousin of America. All of this is like a form of colorful advertising noise signaling that the cherries are ripe for the picking.

All across the circumpolar Boreal forest of the planet the cherry picks up the tab for April flowering to feed and seed a staging ground of staggering proportions for insects, butterflies, and songbirds. These creatures in their long haul of life aim north to a food basket that is sustained by the opening eye of the southern sun.

MEDICINE

The aboriginal medicine of the cherry, *Prunus,* is to be found in the fruit, the leaves, bark, stems, and roots of the small tree.

Science has not characterized the biochemicals of the Boreal cherries, although a few would be expected to be present based simply on guesswork. These would be ursolic acid, a mandelonitrile glucoside of some sort, a ferrocyanide pigment, probably some flavonoids, provitamin A, tannic acids, and various sugars.

A product of injury is also seen in this species that requires characterization of its chemistry. This is a form of almost translucent brown-tinted resin that has antiseptic properties and is protruded as a healing plug for the injured tree. This plug oxidizes into a water-insoluble state in days and loses its viscosity on contact with air.

However, a very close relative of the chokecherry has been used by Chinese doctors to treat tumors since the fifth century. They used the oil of the common apricot, *Prunus armeniaca.* The medical cadre of Mexico used a similar seed extraction for the treatment of cancers. This substance was called Laetrile® and has caused much controversy. However, the basis of its action is of a cyanide release within the cell only when the extract comes in contact with beta-glucuronidase enzymes. These enzymes are produced in growing amounts by the tumerous tissue. The cyanide shuts down the tumerous tissue, leaving the surrounding cells unharmed. This is a strategic chemical bombing of one or more targets. The chemistry that exists within the genus of the *Prunus* species could be expected to follow suit in its action to produce these strategic systems, and the Boreal chokecherry, *P. virginiana*, would be just one of many suppliers.

The aboriginal peoples of the Boreal and various southern nations have commonly used chokecherry in the treatment of colds, flus, and fevers. A cough syrup was made using one bark bundle, which was boiled down in water. Sugar was added to the decoction to sweeten. The syrup was taken by teaspoon.

The chokecherry was also used as a blood purifier, especially for prenatal care. The fresh fruit was eaten. It was also used to clear throats of phlegm for singing or for speaking. It was used to treat high blood pressure. It was part of a complex treatment involving two other ingredients together with the common polypody fern, *Polypodium virginianum,* in the treatment and the management of cholera.

Chokecherry was used as a medicine to stop diarrhea in horses. A decoction of branches, leaves, and berries was used. All members of the *Prunus* group are toxic. These trees produce cyanide-generating chemicals mostly in the form of cyanogenic glycosides. These are more toxic to children with their small body volume and high metabolic rate. The parts of the plant with the highest concentrations of cyanide are the midsummer leaves and their shoots. The next are seed pips or stone fruit. These must be chewed to release the cyanide. That is why the core and seeds of stone fruits are generally thrown away. An exception is the almond, *P. dulcis*. The seeds of the cherries are toxic in fairly small amounts, especially to the young.

ECOFUNCTION

The chokecherry, *P. virginiana*, of the North American Boreal can be seen as five different races spreading out across the continent. Each one is a match for its own special habitat on the landscape. This is seen also in the larger sizes and the darkening colors of the fall fruit crop. The chokecherry, together with the Siberian ground cherry, *P. fruticosa*, the Japanese alpine cherry, *P. nipponica*, and also the Sakhalin cherries, *P.n.* var. *kurilensis* and *P. sargentii* var. *sachalinensis*, are all low shrubby trees. They are in the extreme margins of their global habitat, one that makes their canopy open to damage by browse. These are the species with the most inventive arsenal of cyanide-bearing compounds, all of which are used by these northern trees as antifeeding complexes.

The cherry flowers of the Boreal species are produced very early in the spring, either in late March or early in April. The flowers are produced in long racemes. These begin the process of pollenation from the bottom upward. The flowering itself lasts for a month, into the beginning of May. This is sequential blooming where the tree plans its pollenation to last for a long time. In northern regions this is very important for the tree because all of its eggs are not in one basket. Should temperatures drop, freezing and destroying an early set of flowers, subsequent flowering will allow the plant to set seed to survive. Such survival tactics also match the burgeoning insect population's needs. These Boreal cherries become some of the most important spring feeding trees of the north.

Mature bark of *Prunus serotina*, the black or rum cherry

The cherry is second only to the apple for insect feeding. The cherry produces nectar and pollen, but it also produces another exudate from glandular tissue at the base of the leaves. Wild bees and honeybees visit the cherry for its food. Wasps and ants do also. These insects, for some reason, seek out the chokecherry, in particular the extra nectary exudate, for its springtime offerings.

The crop of red to red-black fruit vanishes rapidly from the trees in the fall. Songbirds, larger birds, and mammals eat the sweetened fruit. The rodent populations collect and eat the seeds. The fruit is also an important bush food of the aboriginal peoples, being very high in niacin of the B complex of vitamins and vitamins A and C. The crushed fruit is sometimes made into cakes as pemmican, or it is occasionally dried whole and stored in this form. It is used as a perfect accompaniment to various meat dishes. These strange sour to bittersweet flavors appear to be part of the ancient cuisine of the peoples of all of the circumpolar Boreal north. They are an acquired taste, such as the mountain cranberry or lingonberry, *Vaccinium vitis-idaea* var. *minus*, of Sweden and Finland. These acidic heath species are the

Cherry

bush foods of northern Europe, filled with regulatory chemicals for health.

The *Prunus* of the Boreal contain a few cyanide-generating biochemicals that are important to butterflies. There are possibly more unknown complexes there, also. One such chemical is amygdalin, which is also called Laetrile®. Another is mande-lonitrile glucoside, or prulaurasin. Of itself the amygdalin is an elegant molecule. It looks, for all the world, like a molecular flock of birds in a wave of flight. This molecule dangles at its outermost edge an offering of cyanide. Both cyanide-containing molecules are toxins. The butterflies use the *Prunus* species that bear these toxins as host plants. The toxins are absorbed by the growing caterpillar and are slated for use as color of the wings, sometimes in mimicry, other times not. The toxin trick protects the butterfly from predation.

Another compound occurs in the *Prunus* that is specifically used by butterflies also in the paint palette of their wings. The complex is ferric ferrocyanide, known as Berlin blue. It is a dark blue color and is part of the pigmentation mixture that is used in mimicry and gender identification for quality sexual service. This, too, is a toxin. Some of the important butterflies that use these toxins are the tiger swallowtail, *Pterourus glaucus*, with its evolution of darker female forms that mimic the nasty, toxic-tasting pipevine swallowtail, *Battus philenor*, of the south; the southern two-tailed tiger swallowtail, *P. multicaudatus*; its two paler cousins, the western tiger, *P. rutulus*, and the pale tiger, *P. eurymedon*; the high-perching Henry's elfin, *Incisalia henrici*; the striped hairstreak, *Satyrium liparops*, a high flyer; the coral hairstreak, *Harkenclenus titus*; and the smart-minded viceroy, *Basilarchia archippus*, whose little life depends on its mimicry of the bigger and better-mottled monarch, *Danaus plexippus*, a floating king that dances in the meadows of North America, ending its footwork in Mexico.

BIOPLAN

The *Prunus* is an anchor group for the Boreal forest and the remainder of the global garden. The chokecherry, *P. virginiana*, is one of these trees that must be bioplanned back into the periphery of forests. A few trees that are carefully planted would become the epicenter of more individuals that would, over time, be dispersed by birds and mammals. Farms, too, should have chokecherries planted within timber set-asides. The beneficial insect ratio for predation would increase, as would the songbird population for the patrol. Hedgerows with existing chokecher-ries should be protected as these areas are sites for the nesting and parental nursing of young birds. Open hedgerows also allow for a greater berry crop in the fall because of wind and root protection. Beekeepers benefit greatly from the bee pasture of all of the *Prunus* species in all of their diversity.

The chokecherry, *P. virginiana*, and its relatives could also be planted in urban parkland as primary or secondary trees. These trees invite the beneficial insect populations along with the songbirds into an otherwise concrete desert. In warm city air, bats compete at dusk for mosquitos and other flying insects that feed on cherries, reducing the need for pesticides. In addition, there is also a cascading effect seen in local city or urban gardens, all of which get groomed by songbirds, thus keeping pathogenic insects at bay.

One very important member of the *Prunus* is getting forgotten in the traffic of modern life. It is the rum or black cherry, *Prunus serotina*. It is a timber tree and one that still decks the halls of major estates of England as scrumptious paneling. These rooms of quiet, restrained elegance were once upon a time, and still are, the marvel of the British Isles. The chocolate milk–colored wood panels were a favorite for libraries and dining rooms where muted discussion was desired. The panels served this noise reduction in spades, giving an atmosphere of sheer enjoyment for the company concerned. The rum cherry was selectively cut out of the forests, and the wood is in short supply. It, too, would be an ideal set-aside cash crop for a farming community. This tree in the past enjoyed the gentle companionship of the chokecherry, *P. virginiana*, a lime lover, too, at its side.

These days the chokecherry does not enjoy a high status, that is, in all nonaboriginal areas outside of the Boreal. This tree has been reduced to the status of a weed tree. This has happened through a loss of knowledge. The pioneers were informed about the chokecherry when they landed in North America. They went on to adopt it for themselves for their many pies, jams, jellies and mixed chutneys that fed their 6,000 calorie per day diet. With the loss of the chokecherry habitat the songbird population is reduced. With this reduction in bird patrol arises an increase of the ratio of insect pathogens.

The removal of anchor species such as the chokecherry from its native habitat, and the treatment of such species as weeds, has serious consequences, especially when these consequences are multiplied by global warming and climate change, circumstances in which pathogens can multiply more quickly just because of the increase of temperature. The tipping point comes with globalization itself, where there are no real barriers for pathogens. They are introduced into new territories and fresh fodder from east to west and from north to south. They will cause pandemics of infection in their path, for birds, for bees, for bats, for trees, and finally for people.

DESIGN

The chokecherries, *Prunus virginiana* and *P. nipponica* var. *kurilensis,* together with *P. sargentii,* are species that will cope with climate change. The design of all of the *Prunus* species is such that they have reflective bark, drought-resistant plumbing, and a leaf structure that has a reflective surface. All of the *Prunus* need a soil with some available calcium in it. The larger trees of the tribe need iron. And in addition to that, they require an oxygenated soil for root health. These needs usually call for light sandy soils with plenty of air pockets. This environment having been found, the tribe looks after its own, with an autumnal leaf fall that has the ability to sheet or coalesce together. This unity of leaves forms a mat that condenses soil moisture like a plastic sheet. This conserves the available water in the soil underneath and recycles it back to the tree's roots in the spring for fresh growth.

The Boreal *Prunus* species should be part of an organic system of gardening or farming. The fresh virginal flowers are members of the *Rosaceae* family. They look and function like single species roses, whose parabolic petals and generous nectary sugar flow matches in volume the butter yellow anthers for pollen.

The anatomy of the flower allows even large insects to use the plants' food sources. These kinds of flowers attract and feed beneficial insects in large numbers. This increase in biodiversity plays an important role in the balance of nature for the organic gardener. It reduces the need for pesticides.

Any of the *Prunus* species should be part of the design of a butterfly garden. The more medicinal species can be part of a physic garden or area within a larger garden. The *Prunus* should always accompany the local native species of a woodland garden or walkway, where native bulbs, tubers, and corms do especially well under the overwintering leaf cover. All of the *Prunus* tribe readily take their part in an edible landscape.

Luckily, the chokecherry, *P. virginiana,* has three very smart-looking cultivars. These have arisen as sports from the various races that occur across the continent. They are all very frost-tolerant and will withstand the cold down to zone 2. The first is an attractive and unusual yellow-berried cultivar called *P.v.* 'Xanthocarpa'. The second is a tomentose form with downy leaves, *P.v.* var. *demissa,* or the western chokecherry. The other is a pretty, conical tree whose fall foliage turns a bright crimson to match the fall crop of berries. It is *P.v.* 'Schubert'. The foliage of the Schubert chokecherry turns a more purple color as the berries disappear.

There is another masterpiece of design for any large country or city garden. It is the glass-faced *P. serrula,* a Tibetan tree. This glamorous small tree has a bark of deep mahogany-red undercolor on which is overlaid a reflective cuticular layer just like a mirror. The tree is vigorous and disease-free with long narrow leaves, unusual for a cherry, but the flowers and tiny cherry fruit remain true to the *Prunus* form.

Regeneration, the hidden secret of all natural forests

Cherry

The pollen bonanza of
the male willow
catkin

Salix

WILLOW

Salicaceae

THE GLOBAL GARDEN

The willow, *Salix*, was the darling of the Old World. It stood firm and was central to the sustainable societies that made up the global family. Few people were unaware of the importance of the willow in their lives. Grandmothers made sure that their grandchildren would recognize the willow in whatever form it happened to be growing locally. The children witnessed the early catkins abuzz with bees, followed by the sleek and elegant leaves that made up the early spring bouquet of the greening canopy. Later, in the shade of the willow, with a screaming dog paddle and a yelp of joy they made their first plunge into the waters the willow protected.

All over the global garden the willow was a source of medicine, dyes, furniture, baskets, wood, charcoal, and agricultural and household items. The willow makes ideal drinking straws and the finest artist's pencils of charcoal. The wood of the willow, because it can take a pounding without splitting, is made into nails, balls, and even cricket bats to win the prestigious "Ashes." The willow tree has its place in religion too, for this tree was used in ritual. The Cree are the last nation to honor the willow, their deer willow, with the sanctity of blessings. When a bear is killed for meat, they weave twenty-one twigs into an offering. This woven piece is burned with a polypore mushroom, *Trametes suaveolens*. This important ceremony of the living of the Boreal is still being performed to thank the spirit of the bear for the gift of the flesh of its own body so that others may continue their lives.

The word *Salix* comes from the Frankish people, a western German tribe that entered the Roman provinces a few hundred years after Christ. They settled in the Netherlands along the IJssel River that feeds into the big bay of Kellemeer. On their flat flooded land the willow grew in luxury. They used it skillfully and were trading partners of the matriarchal Irish, who, in turn, took a dim view of these patrician people. For their glass-ceiling ways, the Irish named them *An Dlí Saileach,* which means the strange ways of the Salix peoples. The Irish may have admired them for their complex baskets but stopped short at the treatment of their womenfolk, a distain down to today in *Saileach* for the Irish-speaking world.

The willows of the global garden are members of the *Salicaceae* family. There are about 300 species, the numbers varying with the many hybrids that pop their heads up in the most unexpected places. Quite often these are to be found in modern nurseries. The species of the willow can vary from tiny creeping alpines that crawl and scratch their way across mountains, to large and noble trees. Of these, surely the golden weeping willow, *Salix babylonica,* is the most painted, beloved, and admired. This native of China was widely cultivated there for millennia as a waterside tree. This tree has been adopted by the gardeners of the world.

There are other willows of great beauty like the coyote willow, *S. exigua,* with tiny silky silvery leaves. This is found in the western portion of the American continent. This willow is found also in Mexico. There is the gorgeous little crack willow of the Welsh countryside, *S. fragilis* 'Decipiens', whose yellow-gray branches have orange-red underbellies.

There are wooly willows of which the common Lapland willow, *S. lapponum,* of Siberia is a shining example because the leaves are cloaked in gray down as a protection against the intense cold. And then there is the Bebb willow, *S. bebbiana,* named for Michael Schuck Bebb, a willow specialist who hunted the tree a century ago.

The Bebb willow is found all across the Boreal forest woodland with another willow, the balsam willow, *S. pyrifolia*. These two willows of the north do as all others do in their tribe: they live near the water. They grow at the edges of swamps, lakes, rivers, and water meadows. They exist as male and female trees, the catkins of both emerging when the first hint of heat hits the frozen bark. Their fuzzy catkins ward off the cold with their excelsior clothing and act as black boxes warming the female flower. This keeps the rich nectar in flow and attracts the flying visitors who have already been lunching on the golden pollen of the male trees nearby.

In the north of the global garden, in the Boreal circumpolar forest, the willow puts the match to the fire of life, and everything else that is living follows in the trail of food the willow offers, be it insect, bug, bee, musk ox, or Lapland reindeer. The willow feeds butterflies and pollinating native bees. It can even add a little taste of flavor to the home brew. But the willow in its travels across the global garden has managed to do something far more important in its life's work. The willow has managed to become an exceedingly important medicinal tree for man and all of the wildlife it touches. This includes the waters of the global garden and even the health and well-being of the fish in their freshwater habitats.

MEDICINE

The medicine of the *Salix* is to be found in the mature bark, roots, and leaves. The medicines of the willow have been known for millennia both in the East and in the West. While the North American aboriginal peoples called the Alabamas and the Houmas were bathing their fevered children in water baths made up from decoctions of the roots and leaves of their nearby willow, the well-fed Greeks were doing the same thing across the world. But they were treating their gouted big toes with soothing bark extracts of the billowy white willow, *Salix alba*. The North Americans also made good use of the black willow, *S. nigra,* a stout gentle giant that inhabits the waterways of the east through Texas down into the rivers of Mexico.

All the species of willow produce their genetically programmed library of salicylates. Some species produce very large quantities of this biochemical complex in their pathways. There are further important medicines waiting to be discovered in the rest of the willow family. Some of the rarer species are under threat of possible extinction by plain ignorance and poor management of waterways because these areas are the historical sites of villages, towns, and cities all over the world. The willows go down as the houses go up.

An accident of rediscovery led the Bayer company to look again at one little biochemical called acetylsalicylic acid that they knew already existed but they had not followed up on. The raw methyl salicylate known as salicylic acid was being used throughout the nineteenth century for the relief of pain, but the gastric distress wrought by the little white pill was worse than enduring the headache. So the company directed its staff to look again to find a substitute for salicylic acid, and they came up with acetylsalicylic acid. And the rest is history. The christening name of aspirin they borrowed from the white-flowering *Spiraea* from which the acid was first isolated.

Today the aboriginal peoples of the circumpolar forest make good use of the willow. These uses are staggering in their complexity, and many are borrowed from the south. These techniques are reworked to fit the northern life of hunting and gathering. Because the willow is strongly antiseptic, antifungal, and antibiotic, the Déné peoples of the north make a powder of the bark of the Bebb willow, *S. bebbiana,* or the balsam willow, *S. pyrifolia*. They use whichever species is at hand. They take the wood and fire it by reduction to charcoal. This charcoal is powdered. The product is kept dry and is used for infected open wounds and running sores.

As with many treatments of the south using other medicinal trees, the willow is a source of treatment both for diarrhea and for constipation. This depends on the method of sectioning of the living bark from the tree. To treat constipation, the bark of the Bebb willow is carefully removed in a slice to expose the inner cambial bark. This is peeled in one long slice upward toward the cutter, and four loose knots are tied into this string of bark. A decoction is made by boiling the four knots, and this is drunk to relieve the constipation.

On the other hand for the treatment of diarrhea, the cut of the inner cambial bark strip is away, downwards from the cutter. This act traps the stronger salicylates that are on their way to the tree's canopy in the phloem tissue of the cambial ring. The four-knot boiled decoction quickly heals the problem of diarrhea.

All the peoples of the Boreal forest use the willow in the usual commercial mode as an antipyretic in the reduction of fevers. It is used as an analgesic in the management of toothache and backache. It is also used as an anti-inflammatory for the treatment of all the conditions of aching limbs that are lumped together as rheumatism.

The root decoction of the Bebb willow, *S. bebbiana,* or the balsam willow, *S. pyrifolia,* is used in the management of the fatigue caused by circulatory dysfunction. In addition a living root section of the two willows, the Bebb and the balsam, is used to stop bleeding and to promote healing without infection. In this case the bark is carefully peeled off the root. This exposes a slippery inner layer of cambium. This layer is excised in one piece and used as a healing antibiotic bandage on the infected area.

Probably one of the most unusual uses for the willow is for the treatment of loneliness. This is for the management of the acute loneliness that is so common in all modern societies, but goes untended by them. Pussy willow, *S. discolor,* is used with white poplar, *Populus alba,* and joe-pye weed, *Eupatorium maculatum.* A bark decoction is made of the willow. This is taken to cause vomiting. Then the total skin of the body is washed with a mixture of the poplar and joe-pye weed, in three separate cleansings. The concept is to rid the outside body of surface toxins and the digestive system also. In this manner the wayward patient is refreshed sufficiently to get back into the saddle of life.

The sweet nectaries of the female willow catkin

ECOFUNCTION

Nowadays the beekeeper is the person most alert to the presence of the willow all over the northern world. In the spring the male and female willow catkins are the first species to flower. This is especially true as the zones get colder. Both catkins emerge on naked branches, and both bear scales to protect the pollen and the pistils of the female flowers. Some have fluff to guard against the cold. This fiber is seen in the common pussy willow. All the catkins produce nectar as a calling card to those on the wing. When the weather is favorable and the sunshine of spring draws on the willow, a good flow of rich nectar is produced. This is the lifeline of all overwintering pollinators who can brave the cold, including honeybees. When the beekeeper sees the daubs of yellow pollen on the landing boards of his hives, he knows that he is in business for another year.

Willow

Bird and water habitat
of the black willow,
Salix nigra

ARBORETUM BOREALIS

Both pollen and the nectars of willow contain salicylate compounds in the food supply. These are antifungal and antibiotic. They are of vital importance as body cleansers for pollinators who have undergone the trials of cold weather and the intestinal problems that come with inactivity. Fowl brood and other catastrophes of the bee world are kept at bay by the medicine of the willow. Healthy pollinators are vital for food production. The pollinators will be more prone to disease as the temperatures climb upward with global warming. The willow in a damp hedgerow or in a nearby riparian area will protect the pollinators. Bees will search an area of 26 km² (10 mi.²) to collect such early medicine for their well-being.

The butterflies of the global garden use the willow in a most extraordinary way. The leaves of the willow produce a salicylate biochemical called salicylsulfonic acid. This is a metal chelating agent. All the butterflies use metals as the electron core in the molecular color of their wings. Metals set in organic ring structures form color. By electron changes butterflies produce the visible and the invisible UV-colored scales of their wings. This pattern of color identifies the species and, above all, formulates the mimicry for disguise, a vital form of protection from predation.

The willows of the global garden live near water. Some live near forests, where their roots can get the benefit of shade, and many more are reduced shrublike species that live out their lives near mountain streams. The willow has learned the adaption of leaf protection in their high-humidity habitat, which would normally leave the entire canopy prone to the fate of mildews. However, an allelochemical is produced called salicylanilide that is a potent antifungal and antimildew agent.

The willow, *Salix,* is the guardian of the watersheds. These species of the circumpolar Boreal forest protect the ponds, the shallow lakes, and the complex wetland systems of the most northern part of the global garden. This series of watersheds represents 30 percent of the forested landmass of the Boreal north. The Bebb willow, *S. bebbiana,* and the balsam willow, *P. pyrifolia,* among other willows of the North American continent act in conjunction with the crack willow, *S. fragilis,* and the Lapland willow, *S. lapponum,* of the European north. These species and all other willow species unite to produce salicylate aerosols as antibiotic, antifungal, and aseptic air cleansers. These are liberated as nonperfumed esters into the air. They are highly water soluble and are added by atmospheric pressure to solubilize into water surfaces. They dissolve into the chemistry of the fresh water as protectants of fish, waterfowl, and underwater life.

Many important fish such as the salmon depend on the natural chemical composition of the watercourses of their habitat. They remember these odors. They draw in this familiar odor-laden water, pumping it over their olfactory nerves near their mouth parts. This trail mix of biological scents lays open the pathway for migration, safety, and successful spawning. Toxins, such as pesticides and heavy metals in the water, eliminate this olfactory memory in a matter of hours.

Labile esters, such as the benzyl ester of salicylic acid, have an action on natural oil fixation and preservation in the body of a fish. Other salicylic acid esters of the *p-tert*-Butylphenyl form serve as light enhancers in a water system. They amplify the electron action of sunlight in the pye electron form of organic aromatic rings. This amplification of northern light into these freshwater systems assists all forms of aquatic life in their fragile habitats. Some water-soluble salicylsulfonic acid complexes are strong surfactant agents. They reduce surface tension and are of help to buoyancy in the sexual life cycle of aquatic species. Of importance also are the salicylic acid bornyl counterirritants that protect the scales of the fish population flexing their mobile surfaces. This keeps the skin of the fish disease-free.

The willow provides high-quality, protein-rich food for fish and aquatic life. The seeds of the willow ripen and are released for dispersal in June, just a few weeks after their early pollination. The seeds are equipped with a cotton corona that helps them float. As they are waterborne the cotton wears off and a fully imbibed seed changes color, turning a bright glistening yellow, the yellow of the salicylate dye. This richly packed food gem has the shape of a tiny fish like a minnow with a fat belly, the corona being the eyepiece and the tail end portion set with its banded tailfin. The whole structure of the seed becomes encapsulated with a gelatinous matrix that makes it easy to swallow and easier still to digest.

Fish and aquatic life have the facility of curiosity. Floating seeds tempt the eye with the glistening golden-yellow color. Willow seeds are like everything else that floats in fresh water: they are up for grabs and the first to arrive gets fed. The early bird gets the worm, while the rest just join in on the queue for feeding time in the Boreal wilderness.

BIOPLAN

The willow, *Salix,* is probably the most widely distributed woody plant in the global garden. There are willows south of the Equator, and willows forge a life for themselves crawling beyond the limits of the tree line of the Boreal. All of the willows have a commission of either air or water protection. In many instances both play an equal part in both elements. There is therefore a willow for all and every habitat that involves moist soils, boggy ground, or rills and rivers with running water.

For farms, private property owners and national parklands, the native willows should be protected and propagated. These in turn will reap a benefit in cleaner and healthier watersheds. Local riparian areas should always have local native willows as important species for consideration. And as luck would have it, there is no easier species to propagate. The seeds of willows have an extraordinary capacity to grow that outstrips almost every species of the plant world. A willow seed will commence germination within thirty-six hours of being liberated from its female catkins on the tree. This happens provided there is sufficient moisture available. This astonishing viability speaks of a heritage of ancient times when levels of moisture were higher and the willow, as a family, thrived. In the past this ardent viability and unfettered growth resulted in rivers teeming with fish and from which potable water was fresh, clean, and drunk in the cupped open hand.

The willows in their fragile habitat of the circumpolar north are presently doing what they have always done, namely protecting the wetland and its species. The willow plays an essential role in the regulation of humidity. This humidity is spilled into the specialized cells of a moss called *Sphagnum*. These cells are called hyaline cells. They can hold up to a thousand times their own volume of water. The superwet *Sphagnum* sits on the surface of the soil like a wet sponge. *Sphagnum* forces the meniscus of water to rise above the level of the ground. The *Sphagnum* grows as one giant green carpet, hungry for every raindrop. These areas form deep *Sphagnum* bogs and *Sphagnum* lakes. It is on this wheel of moisture that the willow, *Salix*, runs.

The black willow, *Salix nigra,* and the white willow, *S. alba,* are both timber-producing trees. The wood they produce is quite uniform in texture. It is light in weight and moderately high in shock resistance. This makes the wood an ideal one for sporting goods such as polo balls, musical instruments, cutting boards, and tables. The wood is also an excellent one for use in the medical field in the manufacture of artificial limbs. In addition, the wood could be used in crash test dummies as human substitutes in high-speed simulations.

The heartwood of the black willow is an attractive light brown going into the deeper nut-brown colors. Quite often the grain holds darker streaks along and in it. The sapwood on the outside of the log is white to a creamy yellow color. This wood, because of its antiseptic salicylate content, is also a medicinal wood, and it is ideal for furniture use where food is handled. Willow wood also makes kitchen cabinets that are lightweight, attractive, and naturally antibiotic producing, ideal for the sustainable home protecting children and pets.

A farm with a riparian area could benefit financially with a bioplanned set-aside of black or white willow as a future source of wood. The trees grow fast and produce the speciality woods for which there is a growing market worldwide.

In Britain willow is widely planted for biomass for fuel for power generation, a process adaptable to North America and the rest of our Boreal regions.

The strange fire of winter willow stems

Willow plantations have been widely used in the past to stabilize sandy banks along rivers and streams. The willow protects the banks from erosion, and the roots act as a mattress binding the shifting sand into the bank. These kinds of plantations are needed in the future to protect vulnerable riparian systems against the flooding effects of violent storm paths that are predicted to come hand in hand with global warming.

In the Canadian Boreal north the Bebb, *S. bebbiana,* and the balsam willow, *S. pyrifolia,* are both used to make a multitude of items from canoe ribs to pipe stems, drumsticks, whistles, rims for birch bark baskets and wicker baskets, and sweat lodge frames. They are also used for stretching hides. The fiber threads of rope, twine, and fishing nets can be made from willow. The willow is also bent and tied into a circle to make the frame for a dream catcher.

DESIGN

Long trails of fluffy white catkins electrify the early spring garden. They seem to touch the ground-hugging crocuses' colors, which appear at the same time. The winter garden is also a speciality of the willow as it descends its branches into cold. The colors of yellow, orange, red, violet, and purple last all winter long as branches in the landscape. These colors become more concentrated if the willow has received a spring pruning the previous year.

There are local willows that will always fit into a larger garden and grace it with a sunshine of yellow in the fall. There are willows for the tiny urban or alpine garden. Of these, the woolly willow, *S. lanata,* is the most attractive. This slow-growing, low shrub bears gray-green, downy leaves and is an excellent choice for a garden with little water. Another is the bearberry willow, *S. uva-ursi,* of eastern North America. This prostrate shrub produces leaves with catkins in the early spring. It is a superb shrub for a scree garden.

For the larger landscape, few trees can beat the weeping willow, *S. babylonica,* for its hanging curtains of grace. This willow has spawned a peculiar sport called the dragon's claw willow, *S.b.* 'Tortuosa'. This small tree carries a more perpendicular growth that winds and twists toward the sky in a most dramatic fashion. Unfortunately it is not very hardy. It survives to zone 4, but only just. A tougher tree for a damp garden, the black willow, *S. nigra,* emerges as a natural for the North American garden. The black bark broken into a rough design is a strong counterfoil for the elegant canopy. This tree can be shaped and managed to fit in with others. It is much neglected where it should be more loved.

The various willows worldwide are slowly but surely creeping into the horticultural trade. Some are extraordinarily beautiful if planted with silver birches, yellow-stemmed dogwoods, snake-barked maples, and mahogany-mirrored Chinese hill cherries, *Prunus serrula.* These all are the active ingredients of the much admired winter garden, where the simplicity of color and form rule the day, with the willow reigning as the fragile queen with its ermine cloak of catkins.

The beauty of the
Boreal, the showy
mountain ash, *Sorbus
decora*

Sorbus

MOUNTAIN ASH

Rosaceae

THE GLOBAL GARDEN

A tree called the *Sorbus* stokes the Boreal with a bittersweet smell. The creamy-white bloom is dusted all over the northern woodland. There are two trees that begin the spring relay across the North American continent. One is the *Sorbus americana* and the other is the *S. decora*. There is very little difference between these trees apart from a larger bloom and flower size. They are, of course, the mountain ash. The bridal passage flecks its way around Lake Winnipeg and the relay stick is picked up by a tougher tree, the mountain-climbing Sitka mountain ash, *S. sitchensis,* that ends the run at the tip of the Pacific Rim.

Across the remainder of the circumpolar Boreal the veil of bloom also appears in southern Greenland with the beautiful showy mountain ash, *S. decora.* Then the European tree, *S. aucuparia,* takes the sprint. This is the rowan tree of northern Europe. It is more timid and will not make the stretch into the searing cold, staying closer to the sun. Another makes the final leap to fame. It is the tree that braves the race because the whales of Sakhalin are leaning on the win. The tree is the Sakhalin sorbus, *S. commixta.* If ever there was a remarkable runner, this is one, a tree of treasure to beat them all.

The *Sorbus* is the mountain ash of North America. Its cousin was christened the rowan tree in Europe a very long time ago. These trees have been for millennia trees of legend and the source of fairy tales for the young and old. They are also major medicinal trees of the Boreal world, and more than one dream was spent there to label these trees as trees of legend. History has tracked these trees throughout time, and the brace of ownership holds the hint of a wonderful ethnic mix because so many northern countries claim the *Sorbus* as their own.

The first pioneers were delighted to find their own rowan tree, *S. aucuparia,* staring down at them when they first landed on the North American shores. They were wrong. The tree was either the showy mountain ash, *S. decora,* or the American mountain ash, *S. americana.* The *Sorbus* was one of the few trees that the pioneers recognized on this continent. It was one of the trees that they had been carefully instructed about back home.

The Scandinavians named the *Sorbus* a long time ago. They named the tree for its crop of red berries. They called the tree *reynir,* for the color of its brilliant fruit. This naming got corrupted down to rowan over time. The Russians and the Irish were excited too. They recognized the *Sorbus* also. These two clannish groups had something else in common that pleased them so much when they bore witness to the rowan berries of the *Sorbus.*

The genetics of the Irish and White Russians were linked a long time ago in the *Tuatha Dé Danann,* People of the goddess, Danii, their ancestors. From this came the homozygous gene of red hair. These two groups called the mountain ash *rua.* This is the ancient name of golden red, a royal color seen in their children's hair that is reflected in the color of the *Sorbus* berries, a tree in which the fairy folk live. Children and fairies go together, except perhaps in Ireland, where the little people of the rowan were a force of darkness whose spells could wreak havoc on a happy home.

The aboriginal peoples of the Boreal took their two *Sorbus* species seriously. The medicine men of the Chipewyan peoples called the tree "the medicine stick" in their language. They used this medicinal tree as part of their pharmacopeia, using its medicines when necessary. They stored the medicinal bundles taken from the tree for use in a large array of important treatments for their sick. The wood of the *Sorbus* has always been considered to have special value for amulets.

But it was the eyes of the Germanic tribes that brought the *Sorbus* to its limits. The peoples that inhabited the northern European forests took note of something unusual going on in this tree. They gathered these observations into their scientific minds and produced a new invention.

The hunters collected the rowan crop. They steam distilled the pulp until they obtained an intensely sticky mess they named *Vogelbeeröl*. This they proceeded to attach to trees with the aid of a stick. The tarlike substance retained its sticky nature. The birds came to call. Their feet got entangled. The hunters were waiting and the birds got transferred into a pie. Such was the life of a peasant poacher who preyed upon the landlord's *Sorbus* while waiting for its other fruits to ripen.

MEDICINE

Both the American mountain ash, *Sorbus americana,* and the showy mountain ash, *S. decora,* are medicine trees of the North American Boreal forest.

The medicinal properties of the Sakhalin mountain ash, *S. commixta,* are unknown. The same holds true for the Kashmir ash, *S. cashmiriana,* with its unusual pink blooms followed by white marble-like heavy fruit.

The medicine of the North American species is to be found in the pollinating flower as a lactone. The young, growing branch tips together with their immature leaves and apical bud are medicinal. The stem, bark, root, and wood also hold strong medicines for the northern aboriginal cultures. The strongest medicines are to be found in the most northerly stretch of the Boreal, perhaps with the exception of the unrecorded Sakhalin species, *S. commixta.*

All over the remainder of the global forest, all of the *Sorbus* species in the upper elevations of mountainous regions represent a genetic library of the unknown. Many of these species are under threat by the activities of war. Most go as yet uninvestigated by science. All *Sorbus* species that dwell in the most marginal of habitats will hold the most potent and variable biochemical load.

A potent, natural, antibiotic is to be found in the two native North American Boreal species of *Sorbus*. This antibiotic is called parasorbic acid. This antibiotic is active against all gram-positive bacteria and protozoa. The presence of parasorbic acid is the underlying basis of the stems, bark, and wood being steam-treated to distill the volatile antibiotic acid, which is inhaled to treat general soreness of the chest with coughs and flus. It is used to clear catarrh and to treat the migraine of headaches. In this steam form the parasorbic acid acts as an irritant, helping to clear these passages for freer breathing.

In a sliding scale of medicinal strength, young *Sorbus* shoots were used to treat ordinary colds. A warm-water tisane was taken to remedy them. Both the stem and the root of *Sorbus* was used as a decoction to treat tuberculosis. Extracts of the ripe fruit mixed with the fresh root were used in pain management for the kidneys, the back, and the relief of the various pains connected with rheumatism.

In addition both stem and root decoctions of *Sorbus* were used as potent fungicides.

The fresh fruit alone was used to promote the process of labor and increase its speed in birthing. The ripe fruit at its peak contains both ursolic acid and ursolic acid-type complexes that play a role in labor.

A decoction of the ripe fruit and root of *Sorbus* was used in the treatment of various cardiac conditions. These aboriginal medicines were antianginal, antiarrhythmic, and antihypertensive. This decoction contained a mixture of the daughter complex of parasorbic acid in its hexanoic form. This is the molecular basis of the *Sorbus* decoction, being a potent beta-adrenergic blocker for the heart.

An interesting sugar is also found in *Sorbus* but it is much more prevalent in the European species, the European mountain ash, *S. aucuparia*. The sugar is called sorbitol. It is being used increasingly as a substitute for glucose in the diabetic diet. This sorbitol has also found its way into chewing gums as a sweetener. This little molecule of sugar spells out an impressive industrial profile. Sorbitol is used in the manufacture of sorbose, of vitamin C, and of propylene glycol, the antifreeze. This nontoxic antifreeze is used in the dairy industry and also in the brewing business, where it is employed to inhibit the fermentation process and the growth of stray fungal molds. Propylene glycol is used to disinfect air through misting. In the food industry it is used as an emulsifier. Sorbitol is unfortunately used in the manufacture of synthetic plasticizers and resins that go into cars, unfortunately for North America because the porcupine population on this continent has developed a taste for the innards of cars, having a sweet tooth from birth.

Sorbitol is even used in printing and in the processing of leather. It is used in the manufacture of cigars and cigarettes and sweetens tobacco. It maintains the flow of ink and prevents crusting in the tips of nibs in all forms of writing. It is added to glycol and to glycerol to lower the temperature of flow in the action of antifreeze for the car markets of colder climates. It is also used as part of the flow dynamics in the ground heating of homes and buildings. It is used in the kitchen to increase the flow of sugar and to keep it from lumping. A similar addition is made to dry shredded coconut and even to peanut butter. Sorbitol is also added as a low-calorie sweetener to many soft drinks and to quite a few table wines. It is used increasingly in diabetic foods that require sweetening since it reduces the nasty aftertaste of many of the other artificial sweeteners. And lastly, in the pharmaceutical world, sorbitol is used as an absorptive agent to increase the digestion of vitamins and other nutrients across the stomach wall into the bloodstream.

The various biochemicals the *Sorbus* has to offer mankind will become increasingly important as climate change progresses. There will be an urgent need for nontoxic fungicides to battle an explosion of fungal pathogens that will arise with acid-laden air that comes from an increased atmospheric burden of carbon dioxide. A portion of this gas on contact with humid air

The flower of the mountain ash, *Sorbus decora,* infuses the Boreal with a bitter-sweet smell in the spring.

turns into carbonic acid. This acid etches paint in very strange places, and even the working radio telescopes will not escape its effects. The acidic conditions will promote all forms of fungal reproduction, especially the higher forms that produce basidiospores and ascospores. These germinating spores, many of which are pathogenic, search out acidic conditions in which they will readily grow and subsequently propagate. The human body, too, as a host, is not excluded from these ancient and enduring pathogens.

In the future the safe storage of many foods like cheeses will become problematic. Fortunately the *Sorbus* species present one answer to this. The berries of the European *Sorbus, S. aucuparia,* manufacture sorbic acid. The European species produces more of this acid than the North American counterparts. Sorbic acid is a potent fungicide that is nontoxic to humans. It can be used to protect foods that need aging, like cheese. All of the *Sorbus* family have the possiblity of a unique production of various fungicidal agents. These have not been investigated and represent a safety net for future food production.

Mountain Ash

Skirting the bound-
aries of the Boreal
region, bloodroot,
*Sanguinaria canaden-
sis*, once in great
demand as a war
paint by aboriginal
warriors

ARBORETUM BOREALIS

ECOFUNCTION

Within the circumpolar Boreal forest the mountain ashes, *Sorbus,* are represented by the showy mountain ash, *S. decora,* the American mountain ash, *S. americana,* the European mountain ash, *S. aucuparia,* and the Sakhalin mountain ash, *S. commixta.* All spread a veil of incredible bloom between the curtain of evergreen trees in the spring. Full canopies gorged with bloom peer out from their growing places laden with the lactones of invitation for the insect world. In the spring these trees are a mass of insect life that makes the fall fruit crop of berries an open house for songbirds on the wings of migration.

The *Sorbus* is in the rose or *Rosaceae* family. Therefore, each tiny flower in the terminal corymb of bloom is the same shape as a single rose. The design is parabolic. The focal center for fertilization carries both nectar and dark pollen. The parabola of petals allows for the flight maneuverability of all insects from great to small within this well of food. Even the clumsy wasps come to feed and can rotate their way out. The rose flowers of the *Sorbus* are the feeding ground for all beneficial insect life within the forest. The native bees come in droves and so does the full range of wasp and moth life. The feeding of these beneficial insects within the Boreal forest system ensures the harmony of a healthy balance of predation within these northern woodlands.

On the other hand, honeybees come to the native North American *Sorbus* only when in dire need. The honeybee is a European insect brought over to the American continent. This feeding bee is better matched to the larger European *Sorbus* with its slightly different pollen grains. Honeybees will both work and fertilize the European mountain ash, *S. aucuparia,* and *S. commixta* if planted in landscapes on this continent. In the Straits of Sakhalin the weather-wrapped Sakhalin or Japanese mountain ash, *S. commixta,* holds the fort for feeding. This uniquely modified species has evolved to produce its nectar and pollen flow despite the turbulence of the spring weather and manages to produce a farmers' market of orange-red berries held staunchly upright for a fall display. The plant is columnar. It holds an armory of sticky bayonets of buds flushed into the wet sky. The leaves have a glabrous surface, too, a waterproofing that never fails. This tree is the fail-safe for beneficial insect predation that makes the Sakhalin Boreal forest a wholesome place, a place that protects the air, the soil, and the waters as a birth bed for the biggest and most fragile mammals of our planetary home.

The fruit of the *Sorbus* is high in sorbitol, a singularly sweet taste that is enhanced by frost. The songbird populations that make use of this tree do so because the sorbitol is a high-energy sugar that helps them in the trial of flight. It feeds the musculature of their bodies with flash-and-dash food that is vital to their survival. Some birds overwinter with these food sources nearby, and many make the dash to breeding grounds afar. The *Sorbus* of the forest woodland is a vital feeding tree for the full cascade of wildlife both now and as the disruptions of climate change begin to occur. The fruit of the *Sorbus* is a medicinal berry helping with the stresses of these other species too.

The story of the life history of the *Sorbus* is of great biological interest. The trees of the *Sorbus* have had to endure a fierce battle for their existence in the global garden. This battle is fought daily on a bacterial stage. The opposition is a bacterium called *Erwinia amylovora.* It is also known as fire blight. This bacterium is to be found on healthy trees in natural epiphytic populations. They grow on the leaves, on the trunks, and on the branches. As the tree encounters stress caused by weather, injury, or some simple miscalculation of crop size, the bacterium becomes a killer. This always begins with a blighted twig that flops, losing its turgor pressure. This is followed by complete dehydration. Then the area shows a blackened scorch mark that is the first visual indication of fire blight.

The battle of fire blight has gone on for multimillennia. Sometimes there is harmony and the tree is healthy. Other times the tree loses, limps on, and dies. The army that holds the fort does it underground. A host of microscopic creatures keep a fine balance for the *Sorbus.* They are called bacteriophages. They survive on bacteria and represent the most numerous species on earth. Their number in water and in soil represents a mind-bog-

Mountain Ash

Liking dry feet for bloom, the small golden slipper orchid, *Cypripedium parviflorum*, bows to a Boreal sun.

gling figure of 10^{31}. These shuttle-type creatures are to be found on the soils around all plants, but more so for the *Sorbus*. Time has presented these particular bacteriophages with the ability to fight fire blight and to hold it at bay.

Bacteriophages and the bacteriophages associated with all *Sorbus* species may represent a completely new library of medicines to handle diseases for mankind. Their mode of action is gene infusion. They would represent a sustainable form of medical treatments without the excreted medicine polluting water. And this is just one of the reasons why biodiversity of the trees of the forest is important; the part is very often greater and more important than the whole.

BIOPLAN

The *Sorbus* is part of the balance of nature that makes the forest itself a sustainable entity. These species are quite often ignored, but they represent an important tree for the northern forest bioplan, a bioplan in which the flowers and berry crop feed beneficial insects and birds that go farther into the forest to perform their life work of beneficial predation.

In northern Europe in the Scandinavian peninsula the fruit of the European mountain ash, *S. aucuparia*, has long been used as a popular preserve. The fruit must be cooked. The red jellylike product is a northern favorite with meat dishes, where the sharp flavor of the preserve is enjoyed. This preserve, like most members of the rose family, contains very small amounts of cyanide, which are necessary for catalytic functions in the human body.

The *Sorbus* comes into its own as a prime tree for erosion control. This tree can withstand considerable drought both in summer and winter. The mature bark reflects light and does not suffer from the damage of southwest injury. This injury occurs during the winter months. It is a common damage caused to the trunk by a difference of temperature. The sunny side melts and expands while the shaded side is still frozen. All this makes the *Sorbus* an ideal tree for exposed sunny slopes that experience a temperature drop between day to night, especially in winter months. In addition the root mass of *Sorbus* is spreading. It appears to be able to withstand periods of intense drought and still maintain moisture in the soil, aided by the dappled shade of the pinnate foliage of the canopy.

The leaves of the *Sorbus* are smooth above and reflect light. They are hairy underneath and can maintain and capture rising ground humidity to maintain photosynthesis under the toughest conditions. This benefit is seen in the tree as it matures into its teenage years, because once a crop of fruit has formed the tree will continue faithfully to produce an annual crop of berries thereafter. So the tree, in its mountainous habitat, will be a dependable food reserve for wildlife when all else fails.

The lactone calling card of the antibiotic parasorbic acid adds to the health vector of the Boreal forest. As this biochemical becomes airborne into the atmosphere in the springtime, it acts as an air scrubber that travels far and wide. The benefits are felt in the remainder of the global garden as clean fresh air.

All of the *Sorbus* species produce a very strong fine-grained wood that is also fragrant. This wood is of itself a medicinal wood. In the past axe handles and other small household items were made of this pale brown wood. It gasses off health-giving vapors, and some imaginative cabinetry could transform this wood into a new sustainable industry for household cabinets. *Sorbus* wood could well replace many plastics with their toxic fumes of airborne monomers.

DESIGN

The devil is not in the detail for this tree that was once used for exorcisms. Rather, the *Sorbus* family of trees are small and contained. That is why they have been so popular as specimen trees for both city and country gardens of North America. The tree with its pinnate foliage and delightful bark is beautiful all year round, with its blossom of spring and its fall canopy of orange-red berries.

Of the North American mountain ashes, the showy mountain ash, *S. decora,* is probably the best medium-sized tree with the largest blooms. This tree has produced a smaller fastigiate sport of unknown origin called *S.d.* 'Nana'. This little tree, despite its reduced size, is not as hardy as its parent, being only able to withstand temperatures up to zone 5. However, for a small garden this tree has the distinct advantage of being very slow growing.

It is hard to beat the American mountain ash, *S. americana,* for a larger country garden or parkland. In the more open space of a lawn this tree changes into a tall, noble specimen with tight, vigorous growth. This *Sorbus* if planted near a vegetable garden will entice an ample number of beneficial insects to pollinate the vegetables of the garden. It will then follow this feat by attracting an army of songbirds that will go into patrol behavior for pathogenic insects in the immediate garden area.

The European species, *S. aucuparia,* is not tolerant of alkaline soils. On an alkaline soil, the tree is short-lived. It is also not as cold-tolerant as its American relatives. But breeding has produced many fine cultivars, one of which is extremely beautiful, again for a small garden or the restricted area of a water garden. This is *S. aucuparia* 'Pendula'. Another charming fall cultivar is the award-winning *S.a.* 'Fructu Luteo', a fine little tree with amber-yellow fruits. There is also a *S.a.* 'Laciniata', with deeply cut fern-type foliage for the more elegant style of an urban garden.

Mountain Ash

A brown algae, *Fucus vesiculosus*, is a serious seashore protector of the entire Boreal world.

ARBORETUM BOREALIS

MEDICINAL PLANTS

Agastache

HYSSOP

Labiatae

A giant plant reaches for the Boreal sky. It is hyssop, *Agastache foeniculum.* This towering creature is a member of a very old family of the global garden, one famous for its strange array of medicines. It is the *Labiatae,* a family recognized by its boxed square stems.

Hyssop has been known for a very long time. The common name is ancient and can be traced to its Semite origins in Hebrew, *ēzōbh.* Indeed the common name has traveled time with very little change, except for the bump in the Bible. The hyssop herb of the Old Testament could also have been a marjoram, a fragrant herb of the dry and rocky grounds of Palestine.

Hyssop comes into flower in the early fall. The plant seems to grow overnight from a barren ground. The flower is a whorl of blue-violet tiny florets. And the inflorescence itself, if stroked gently, produces the fine fragrance for which hyssop is world famous. The inflorescence, its sepals and terminal leaves, is covered with very fine glandular secretory hairs. These get broken easily with touch and release the volatile oil of hyssop into the air as a lingering, delightful fragrance.

The world of the insect is not immune to the flowers of hyssop. They come at a high gallop, especially the bumblebees and smaller native bees, arriving in droves, helping themselves to the pollen and the sweet nectar that is pumped from the base of the funnel-shaped corolla. The honeybee can manage this floret, too, as can the very large swallowtail butterflies.

Hyssop has been used as a culinary herb for a very long time. The aboriginal world of the Boreal west makes good use of the hyssop to flavor teas. The mature leaves are harvested, air-dried, and stored. They are boiled to make tea, and sometimes the mature leaves of hyssop are added to other teas to increase flavor.

Hyssop, *Agastache,* of the Boreal reduces perspiration.

Hyssop

Hyssop has been used all over the world to treat coughs, colds, and bronchitis. The herb has an excellent effect on the stomach, increasing gastric secretions. Strangely, the herb reduces perspiration. The flower head is a first rate breath freshener.

Allium

ONION

Alliaceae

The spring leaves of the nodding onion, *Allium cernuum*

Onions will make you cry. The common table onion is in the *Alliaceae* family. The onion, the chive, and the bulb of garlic are related because they cause the same crying game with the gunpowder of sulphur.

The Boreal bends to these table missiles that stalk the virgin forests of the north. It features the common chive, *Allium schoenoprasum,* which takes a shorter form for Siberia as the Siberian chive, *A. sibirica.* There is the more commonly known *Allium* of the south, the nodding onion, *A. cernuum.* The prairie species, *A. stellatum* and *A. textile,* drive a sledge into the midwestern Boreal arena.

When the global forests were intact, the *Allium* was a species to be reckoned with on the forest floor. They fitted their bulbous bodies in between roots and sucked the heat of the reflected sun from the trunks of trees into their earthly homes. This heat caused the placental plate of the *Allium* to grow its little brush of ivory white roots into the fall. This anchored the species to produce the lily-type flower in multiples of three for pollination. And from the locular membrane rolled the ball bearings of jet black seeds. These hid themselves under the layers of detritis of the forest floor, for there would be no end to the *Allium* as long as the virgin forest lived.

The *Allium* is to be found recorded in the works of the Chaldeans of Babylonia, the Egyptians, and the Greeks. Two millennia ago, Pliny of Rome recorded in detail the varieties to be used for cultivation. The peoples of the Boreal still stalk theirs from the wild.

The *Allium* changed one essential amino acid called cysteine. The result is the compound alliin. This little devil has a strong smell of garlic that pulls out the tears. The sulphur is released by an enzyme, alliinase, which is activated when the bulb is crushed. Allicin and other compounds are left behind. These are antibacterial and antifungal. Extracts are used in the management of hypertension and the true killer, arteriosclerosis. These compounds also reduce blood pressure and blood sugar. They protect against the common cold. Not bad innings for a tear-jerker . . .

Apocynum

SPREADING DOGBANE

Apocynaceae

A butterfly called the monarch has something in common with Africa. The monarch is the fabled black and burned-orange butterfly that soars in a sacred migration the astonishing length of the North American continent. This ritual of passage is an annual one. This journey is a life cycle. It is undertaken as a sexual rite of survival involving the release of sexual scents to subdue the more tremulous female butterfly.

As spring ripples up the face of the North American continent, the solar force of the sun feeds the milkweed family. The species that flower with the meniscus of the sun's travels are the milkweeds, *Asclepias,* and the dogbanes, *Apocynum.* These are ancient species of wild, open, wind-blown gardens of the natural world. They have a provenance that matches life itself. The milkweeds and dogbanes in the billions play host to the rigid requirements of the traveling monarch's needs. These speak to the butterflies' remarkable life and survival.

Millions of years ago the milkweeds and the dogbanes made a connection with the monarch. This was understood by the plants and the butterfly in their own realities. The plants changed their internal sap into latex, a soluble form of rubber. The plants would sacrifice some leaves for the monarch's food and the latex would seal the wounds. Inside in the latex a biochemical broth was concocted. The broth was toxic. It was a clever, water-soluble mixture of poisons the likes of which are a wonder in themselves.

In the dogbane there is a glycoside complex that speeds up the pace of the beating heart. It revs up the heart to such a pace that it can stop beating. The monarch feeds on the dogbane, and these toxins do not harm the butterfly. But the toxin load in the insect is high, so high that it will harm the predator who has his

The monarch butterfly uses dogbane, *Apocynum,* as a protective poison.

beady avian eye on the colorful flying thing. This toxin protects the life of the monarch. Other butterflies even mimic its colorful pattern in order to hide in its harmony.

The common bond across the world is in eastern and central Africa. Tribesmen since time immemorial used an identical toxin trick on their arrowheads to kill their prey. And in the Boreal forest of North America, spreading dogbane extends the home of the monarch up and into the crown of the world, a suitable site before setting sail for the pines and cypress of the south.

Aralia

WILD SARSAPARILLA

Araliaceae

Wild sarsaparilla,
Aralia nudicaulis, the
ginseng of the north

Wild Sarsaparilla

There is a very remarkable plant in the lower regions of the North American Boreal forest. It is wild sarsaparilla, *Aralia nudicaulis*. It is a member of the ginseng family, *Araliaceae*, which has been made famous worldwide by the remarkable attributes of one of its members, ginseng, *Panax quinquefolia*.

Ginseng has been used as a panacea. Its wilder relative, sarsaparilla, has also been used for that purpose in the Boreal north. The notion of a panacea is very old. The concept has forged its way through folklore, through the dreamscapes of the medicine men and women, through to the science of today. A panacea is a naturally occurring stimulant, usually from the plant world, that can fine-tune the carburetors of the machinery of the human body. They can fine-tune an ailing body back to health and they can rev up the engine of health itself. There are very few plants that can do this, and many are extinct. One is still to be found in the Boreal, growing in the odd pockets of rich deciduous woodland. It occurs in soils that are higher and more well drained. Such soils hold that peculiar, crumbly, sandy mixture of oxygen-rich pores that can protect a growing rhizome, allowing it to expand and offer up the harvest of its exquisite medicines.

Chinese, Japanese, and Russian scientists have shown the biochemicals responsible for the panacea effect to act as a stimulant for the midbrain, heart, and its blood vessels. They act on the function of the heart as a muscular pump and for the general overall metabolism of the body. They act as a stimulant for all internal secretions of the body, too. They fire up the central nervous system into a more reactive state, and they lower blood sugar. Some or all of these panacea effects would be found in wild sarsaparilla, *A. nudicaulis*.

Species with a similar chemistry include *Smilax medica, S. ornata*, and *S. leucophylla*, which are standby medicines of tropical America and the Philippines. In the Boreal forest system of North America wild sarsaparilla, *A. nudicaulis*, will go the way of the dodo and bison before we will have fully realized what we have all lost.

Cornus

BUNCHBERRY

Cornaceae

The twin wonders of the North American continent, the timid passenger pigeon, *Ectopistes migratorius,* and the little red berry that fed them, have almost gone. The bird is extinct and the berry may soon go too.

Once upon a time, not so long ago, the airstreams of North America were filled with clouds of pigeons that darkened the sun. They flew catching the air currents with a lift that broke them into cresting waves. Their sounds were feather soft, succinct with the clicks of a million wings, always caught in a banded move to nest for spring and to survive for winter. The story is still here in the oral links of memory. Those passenger pigeons who dazzled mankind were there with the lyric of their pulsing life. And man. . . . his feet are still in the clay.

The little red berry that fed the legions is called bunchberry or *Cornus canadensis.* It is called dwarf cornel, dwarf dogwood, crackerberry, and puddingberry, too. It is one of two perennial herbs in a strange little family of smaller shrubs and trees called the dogwood family, *Cornaceae.* The flower flares four white bracts to look pretty with a hidden set of tiny green petals inside. These give rise to a bright red scarlet set of fruits called drupes. They are edible and bland, but they will explode with seed when the temptation arises. After a span of clear days, the deck is set for a catapult of power to spread the tiny seeds. The plant may be lowly, but its gifts are not.

The bunchberry exists in the billions in the circumpolar Boreal of North America, Greenland, Europe, and northern Asia. The color of the berry sends a hidden message to the avian eye. The message is like a rainbow. It begins with color, the color red of ripe fruit. This color is made up of two very closely related

molecules with very different abilities. One, pelargonin, has the capacity to transcend the old problem of mixing oil with water. Pelargonin is hydrotrophic. It makes water more available to act biochemically. The other, cyanin, affects the avian eye and improves its vision to see at night. The cyanin bonds with different sugars to make this night vision more powerful. In different fading lights the cyanin changes. But when it is eaten as food, it changes the eye also. The little red berry banishes night blindness, and what's more, the passenger pigeons must have known it too!

Bunchberry, *Cornus canadensis,* fed the now-extinct passenger pigeon.

117

Bunchberry

Empetrum

CROWBERRY

Empetraceae

A black pearl called the crowberry, *Empetrum nigrum,* calls the margins of life, home. Sitting between land and sea, this prostrate species spins a mat of the deepest green. For the tiny purple spring flowers, flung with salt spray and bitter cold, the song of life endures. Each flower explodes into a magnificent black fruit that pulls in the heat of the summer's sun to make its complex sugars darken down. Each fruit carries the quiet, still luster of invitation.

The crowberry is awash on the peaty shores of the circumpolar world, tracking its way around all the edges of the seas into Iceland, Greenland, Ireland, England, northern Europe and across into Asia and back again into Alaska. The species creeps out of acid bogs and measures the fetch of the tundra for growth. It is found on moraines, in between glacial rocks, and in diminutive closed valleys of growth in wild frontiers of sweet and savage air.

The crowberry of the Boreal is one of the most important migratory bird foods on the planet. The species inhabits the major staging grounds in flight and supplies much-needed food that is high in minerals, carbohydrates, and antistress vitamins. These supply the energy needed to cross over large bodies of water or tundra. Birds of passage require this high-calorific food as a hedge against fatigue. They also require the rhamnose sugar complexes to give a greater acuity to their vision in their travels through their north-south corridors.

Despite the fact that the crowberry has lived on the margins in a low-pH situation similar to many important drug-producing plants, this species has not been characterized for modern medicine. The Cree have used the fresh dichotomous branches as an emergency medicine to reduce the elevated temperature of fevers. They simply chewed the green leaves until the temperature went down. Their children were treated with a decoction. The crowberry gave the Inuit their astounding health and endurance in the past. And the crowberry, today, together with its wild yeast bloom, supplies a new and growing industry of alcoholic beverages to while away the midnight hours of many Icelandic homes.

Crowberry

Epilobium

FIREWEED, WILLOW HERB

Onagraceae

Fireweed, or *Epilobium angustifolium,* and river beauty, *E. latifolium,* are members of the medicinal family *Onagraceae.* These plants are found all over Siberia and northern Asia. They are drunk in Russia as a tea called *Kapporie,* or Kapor tea, a gentle stimulant for a rough climate.

Fireweed has a history of its own. This plant smells fire and loves to spread its wings into open spaces. So it probably was with some astonishment that those who remained to live after the bombing of London during the Second World War were witness to an extraordinary new garden. Fireweed began to grow and flower out of every nook and cranny. It was seen edging its way upward between the shattered pieces of pavement and the ruins of craters the doodlebug bombs had made. The doodlebug bombs were designed for concussion, often bringing fire in their wake. In the case of London, though, more than fire followed the incendiary path. Fireweed too, came to begin a garden anew.

America also had its taste of fireweed. It is long forgotten now. The First World War needed timber and American forests came down. Fireweed went berserk in the new open spaces. The seeds dehissed in such numbers that the fly screens of porches and meat safes got clogged. Food rotted while housewives worked desperately to clean up.

Fireweed, *E. angustifolium,* and its more nutritious sister, river beauty, *E. latifolium,* are medicines that are as old as the hills. Both plants are edible, making excellent salad greens. For their medicine, the plants are harvested at the flowering stage. This is at a time when the basal region of the stems have hardened. At this time the roots are mature, too. They bulge slightly above the soil surface.

Fireweed and river beauty contain myricetin, which is a

River beauty, *Epilobium latifolium,* makes a popular tea for the Russian north called Kapor tea.

flavonoid. This is also found in sperm whale oil and is the narcotic analgesic portion of nutmeg. These compounds are used industrially as softeners. Taken as a medicine, they maintain the health of the prostate. Fireweed and river beauty help to reduce the benign tumors of this area. They regulate by releasing a number of prostaglandins . . . all to be found in the *Epilobium* lettuce leaves of the Boreal north.

119

Fireweed, Willow Herb

Equisetum

HORSETAIL

Equisetaceae

A dinosaur dish,
horsetail, *Equisetum,*
of past millennia

There are thirty-five horsetail, *Equisetum,* species worldwide. They are missing from Australia and New Zealand. In every culture, old and new, they were called horsetails. The plant looks like a green horse's tail with all of the tough strength of horse's hair. The whiplike action is the same, rough and hard.

Equiseta were an important international trade item of the entire Middle Ages. Every single wooden house and pewter plate was polished with the sandpaper of common horsetail, *Equisetum arvense.* This precious item was part of the metalsmith's shop and the carpenter's world. It was also used in the apothecary. There it was labeled in Latin as *Cauda equina.*

Across the Atlantic, *Equisetum* was used by the aboriginal nations all across the North American continent. For some it was a food, carefully prepared because of its toxic load of silicic acids. For others it was a medicine where the fresh juice of the vegetative stems and the dried sterile vegetative stalks were used as a diuretic, as a compound to reduce swellings, and as a geneto-urinary astringent. It was also reduced to an ash and used as an antiseptic and to stop bleeding.

In the history of the world, horsetails take an important place at the table. They were the ancestors of the global forests of today. Without the entire order of *Equisitales* we would not have had the oxygen to evolve into the species of our own class of man, *Homo sapiens.* In our species of man, the female carries an embryo *in utero* for a full nine months. This heroic task requires a plentiful supply of oxygen to the serving placenta that feeds the growing child. Oxygen is the metabolic energy that fires the healthy growth of the unborn child. Without the history of trees, aerating forests and the invisible forests of the ocean, this would not happen. The fragile process of human life would be impossible.

In the hot and humid Carboniferous period of 360 million years ago, the world flora was different. *Equiseta* grew into forests that can be seen today in coal mines as fossils. These prehistoric trees are known as calamites. They were huge and green. They photosynthesized, snapping the carbon load out of a deadly carbon dioxide sky. They used specialized air tanks to breathe; they were in the trunk and the long air canals of the ridges of the cortex. *Equiseta* gave the planet a kiss of life a long time ago.

Fragaria

STRAWBERRY

Rosaceae

The fruit of the strawberry, *Fragaria,* is sweet and it is fragrant. The strawberry was a favorite of the Romans, who named it for its fragrance, calling it, *Fraga.* The common strawberry, *Fragaria virginiana,* is seen all over the circumpolar forest from east to west. Occasionally a taller version raises its head above a pillow of rushes. This is its bigger sister, *Fragaria vesca,* the sow-teat or woodland strawberry. It too roves the Boreal world.

The aboriginal peoples of North America have had a love affair with the strawberry since time immemorial. They have treasured this fruit for its sweet taste and medicine. The Cree of the Boreal called it *otahimin,* or heart berry, while the Chipewyan called it *ídzíaze,* or little heart. The Mohawk of the south give a special acknowledgment to the strawberry as a leader of all plants.

Even the highwaymen of the Middle Ages in Europe blessed this fruit because it was their ready source of cash. The rich, sitting in their moated castles, had a thirst for the little wildwood strawberry to finish off their feasts. The picking and transportation of the wild crop was by foot or horse. The price of the strawberries was sky high. And so was the highwayman's haul, in golden guineas.

The leaves of the sow-teat strawberry, *F. vesca,* contain ellagitannins, flavonoids, including leucoanthocyanins, sugars, ascorbic acid, mucilage, and essential oils. The remaining eleven species of strawberry have not been investigated. They occur from Chile to California to South Africa.

In the Boreal the leaves, the roots, or the entire plant of the common strawberry, *F. virginiana,* was boiled for a decoction for forty-five minutes. The result was cooled, strained, and

Highwaymen of the Middle Ages stole baskets of wild strawberries, *Fragaria.*

drunk as a decoction to treat various heart conditions. The roots, leaves, and over-ground runners of the same plant were used as a decoction in the treatment of diarrhea.

Elsewhere in the world the English had a custom to use a gargle made from the roots and leaves of the sow-teat strawberry, *F. vesca,* to heal periodontal disease and to reanchor loose teeth. South Africa's aboriginal population used the Cape strawberry, *F. capensis,* as a root decoction as a fungicide to treat thrush, a yeast infection of the mouth.

Gaultheria hispidula

WINTERGREEN

Ericaceae

Wintergreen, *Gaultheria procumbens*, is a very old aboriginal medicine.

Alcohol and tobacco, the lax-legged twins of civilization, are followed closely by another drug. It is aspirin. The United States alone gobbles up around 50 million pills on a daily basis. This is all the result of a single-handed action of a certain Dr. Gaultier who practiced in Quebec in the year of 1750.

The good doctor had listened to the local medicine man. The winter had been hard. Coughs, colds, flus, and chest problems had exhausted him on his daily rounds. He couldn't seem to get a grip on the situation. One minute his patients seemed to be on the road to recovery only to slip off it into another set of bed-drenching temperatures. The medicine man arrived with a hay bale of wintergreen, *Gaultheria procumbens*, under his arm. There was also some creeping snowberry, *G. hispidula*. There was less of that. It was rare.

The tea did the trick. Wintergreen lowered the temperatures immediately. An infusion of the leaves was gargled. A warm-water douche was used for the headaches. The aches and pains that pierced the joints and limbs were settled down with a stronger mature leaf decoction. The oil was able to penetrate the skin and jump at the pain *in situ*. Immediately, wintergreen became the new wonder drug. Later, it was picked up by the Bayer company. The pills are pushed today, but the plant is better than the pill.

Wintergreen, *Gaultheria*, is part of a very old family that can be seen in New Zealand, Tasmania, Australia, and South America. It is one of the important health species of North America. It is found in China, Japan, and on the lower stretches of the Himalayas into Tibet. In the Boreal it is part of the conifer forest system in wet woodland and muskegs. In this acidic area, creeping snowberry, *G. hispidula*, survives as a dwarf form in which all the medicines, too, are concentrated. This species of *Gaultheria* is rare in the Boreal. It is really a subshrub that has learned to survive the killing effects of grazing and bitter cold.

A leaf decoction of creeping snowberry, *G. hispidula*, is used to treat high blood pressure by the Cree. It is not known what else this miniwonder will do.

Gentiana

GENTIAN

Gentianaceae

The autumnal wildwood of the Boreal produces a flower of astonishing beauty. It is the gentian, a plant that stalks the moist meadows of the north and blooms with a magic of midnight blue. The flower is called the bottle gentian, *G. clausa*, a plant of mystery and medicine.

There are 350 members of the *Gentianaceae* family. They are mostly perennials that love the north temperate world, some preferring mountains and others the cold. But all adore that tincture of moisture at their roots. They are true species of the wild. They resist capture. Many a glasshouse has been built by lords and ladies to herald their fleeting beauty, but money cannot buy the essentials of nature. Especially not in a cage.

There were many gentians known to history. The most famous one was the yellow *G. lutea*, beloved of the Greeks and named for their king Gentius. He drank it as a potent gastrointestinal tonic. This plant has been used for two millennia. It is found today as the bitters of vermouth. The blue gentian, *G. clausa*, of the Boreal has been used as a medicine in North America since time immemorial. The gentians appear to have a common bitter glucoside called gentiopicrin that also has a strong antimalarial action. The other major component is gentisic acid, an analgesic and anti-inflammatory. This is the basis of the aboriginal use of the bottle gentian, *G. clausa*.

As a painkiller, the roots of three plants were steeped in 250 ml (1 cup) of water. This was used in the form of a spray into the mouth, or it was placed into the eyes, drop by drop. This treatment was used for headache and general pains of the eyes. A stronger decoction was used for muscular soreness. In addition to the internal dosage the pounded roots were used as a surface poultice.

The Boreal gentian, *Gentiana clausa*, is a plant of mystery and medicine.

All of the 350 gentians in the global garden contain antimalarial medicine of varying degrees of efficacy. One by one these plants will die off because of climate change, for their weak link is available moisture. They cannot be captured. They represent all that is wild; maybe even all that is true.

THE TINDER FUNGUS

Hymenochaetaceae

A sacred plant, the tinder fungus, *Inonotus obliquus,* of the Boreal

There are sacred places and sacred things. There are even sacred names. The old Celtic means of naming of the mother of Jesus was sacred. She was Mary, but her Gaelic name *Múire* was used for her alone. This name is in the old Irish stone churches, and a child is never given that same sacred form. A female child can be called Máire but never, ever Múire. Such things are unthinkable.

There are sacred plants, too. Many are truly ancient implements of some divine ritual. One of them is the tinder fungus of the Boreal. It is the divine plant of the north.

For the aboriginal, divinity in the dream of a plant leads them into the past and sustains them for the future by teaching at the present moment. This notion of the present moment has been sacrosanct in religion for a long time. But it has been sacred to mathematics, too, beginning the journey with zero and ending with infinity. It is the poem, the divine dream of the universe.

The tinder fungus has a global passport. It is found and beloved as part of the circumpolar Boreal forest. It also grows, strangely enough, in the tropical heat of Ceylon and in the stunningly different forest structure of Australia. This fungus has relatives in Borneo, Central Africa, and tropical America. The mycelium in all of these countries does the same thing. The fungus lands as a roving ascospore baby that produces hyphae. These hyphal threads penetrate the bark of host trees and then get under it, lifting it up for further exploration. Next the act of wood cannibalization comes. The devouring of the wood lignins feeds the fungus while it grows to kill the tree. This frightful feeding forces the fungus into a marriage that produces a next generation of ascus, inside which are eight infantile ascospores. These start the cycle of feeding all over again.

The tinder fungus is of the *Hymenochaetaceae* family and goes by the name of *Poria obliqua,* or *Inonotus obliquus* for other audiences. It is black. It is crusty. It has an orange central mass. And it is loaded to the hilt with a constellation of life-altering medicines, one of which is the ageold ergot complex, the bawdy house beast of *Claviceps* and *Cordiceps.* These are two fungi that rerouted the Middle Ages.

The tinder fungus is handled by medicine men. It is harvested only by them. The dry heart of the fungus is crumpled. It is burned and the sweet-smelling smoke is inhaled as medicine. This medicine is hallucinogenic. It focuses the internal dream for divination, one ancient pipeline to the wisdom of the gods.

Ledum

LABRADOR TEA

Ericaceae

The sweat lodge is an ancient medicinal tool of the aboriginal peoples of North America. Its invention was refined in the Americas from the wisdom of oral knowledge. It was based on visions of the medicine men and women who were told how to handle this cleansing technique.

The world of the person is a trinity that is made up of the external, internal, and spiritual portions of the body. All of these regions must work in harmony to produce healthy living. Each has an equal share in importance. The share in the triad of the spirit is as important as a functioning liver. The genius of the sweat lodge mends all three.

One of the most powerful medicines in the northern world was used in the sweat lodge. The species is called *Ledum*. It is represented by four members of the circumpolar arctic world, into northern Asia and many parts of Siberia. It is seen in China, Japan, and the rosary of the Kuril Islands. The species more common to Canada and Greenland is Labrador tea, *L. groenlandicum*. The *L. palustre* is more often seen across the Bering Sea. Its dwarf sports, *L.p.* var. *dilatatum* and *L.p.* var. *decumbens,* hold the greatest medicine load. All four species hold differing essential oils of extraordinarily mobile character in the human body.

All of the *Ledum* species grow in strongly acidic soils. They enjoy their water near their roots like many of the rest of their tribe, the heath or *Ericaceae* family. The mature leaves are more commonly used for medicines although occasionally the entire plant is used. The evergreen leaves are small and folded backward to protect the mat of brown tomentosum and glandular tissue underneath. The upper surface acts as a water-repelling roof to the water-soluble, vulnerable medicines below.

Labrador tea, *Ledum,* is a spiritual medicine of the aboriginal peoples.

A tea, made out of the leaves, is dropped on the heated stones of the sweat lodge system. All of the mobile aromatics are then released. Occasionally, for bad breath, a leaf is chewed. A stronger decoction is used to treat heart and kidney ailments. Whole-leaf applications are used for cracked nipples, an increasing condition that stops breast-feeding. The dried and powdered leaf is added to oil to treat all manner of skin conditions. The whole plant decoction amends hair loss.

One of the chemicals of *Ledum* is arbutin. It is a diuretic and a urinary anti-infective. This has a firewall of right-handed oligomers made up of catechin tannins. There are many more, like the whirligig ledol, the essential oil.

Labrador tea or *Ledum* can be toxic in large amounts. Its honey is also toxic. The idea is to treat it like morphine . . . with a great deal of respect.

Lonicera

HONEYSUCKLE

Caprifoliaceae

The honeysuckle, *Lonicera*, has a medicine that makes hair grow.

126

Honeysuckle

The honeysuckle, *Lonicera*, is just another boring business for a hungry bumblebee. The nectaries are down on the floor of the gynecological layer at the very base of the petals. They are too far away for the bee to reach. They are calling to its hunger with a ripe sweet smell. Undaunted, it drills a tiny hole right through the petal wall and feeds to fill its desire. Then the others come running; the frenzy ends with the click of the hummingbird. Occasionally, there is one more, a child. Small fingers pluck the flower and the base is sucked like a lollipop. Such are the hidden delights of childhood.

The honeysuckle vine, which is commonly called the woodbine, has been used for its medicines since well before the twelfth century in Europe. There the twining deciduous flowering vine was harvested either at or after flowering. There it was used as an antiseptic and as a diuretic. Now it is rarely used outside of the flower border.

The honeysuckle occurs all over the world. It is represented by 150 species. The late-flowering, hardy vine of the Boreal, *L. dioica*, is just one of many. The vine gets dispersed by birds. The toxic berries are a delight to them. The seeds are scarified by the passage through the bird's entrails. The seeds swell. They begin to grow.

The chemicals of twining rule the honeysuckle, clockwise for North America and possibly counterclockwise for the southern hemisphere. These chemicals are light sensors, similar in the pathways of all living things. These chemicals make up part of the serotonin pathway complexes. These indole alkaloids are found in high concentrations and make the day for the medicines of this species. They are the precursors of light sensing, which is important for all climbing species to make way for the sun so that flowering and fruiting are accomplished for survival of the species.

In North America, the honeysuckle vine represented an ancient food for the aboriginal peoples, some of whom did not know table salt or pure sugar. The mature stems were peeled, nibbled, and sucked for their sweetness.

In the circumpolar Boreal the honeysuckle was used in many ways. The Cree used mountain honeysuckle, *Lonicera dioica*, to make the hair grow longer. They named it the "long-hair plant." The mature stems were soaked in water. Then this water was used as a hair rinse to make the hair grow, the water-soluble indole compounds acting directly on the scalp.

Mertensia

BLUEBELLS

Boraginaceae

The king's plant is found in the Boreal. The Cree named it a long time ago. This beauty is called *Mertensia paniculata* var. *borealis*. It is the bluebell of the Boreal and if ever the Cree nation had an item for barter with the rest of the world, this is it.

The bluebell's flowering stipe hangs with a carillon of bluebells. The corolla sometimes begins as a pale pink color when young. As it ages into one month of flower, the corolla turns a baby blue. Occasionally there are streaks of white in the odd petal. The calling card for the insect world is a mild perfume that floats on the breeze. The powdery smell begins with the dampness of the dew and drives itself around the ripening flowers with a flush of sweet delight. The flowers are pollinated by dark native North American bees. They come until the nectar is no more.

The bluebell vanishes back into the ground after flowering in May. The Boreal bluebell begins as a small dark wrinkled seed a little smaller than an apple pip. The seeds are expelled by the plant in the summer and lie on the surface of the dark, rich soil in complete disguise. In the first days of spring, the seed germinates. From this birthing a small green flagship of one solitary leaf appears from the plumule. This continues to grow all spring, ending in late May. The flag goes down and the tiny tear-drop rhizome is just beneath the surface of the soil.

The following spring the flagship begins again. This time the lanceolate leaf is larger. The dark skin of the rhizome continues its ovoid shape, moving into a fat pear form. This rhizome reaches down and into the soil deeper for its own winter protection. The third year the bluebells are seen. They are a glory of the northern spring. They speak only of a lush beauty.

Somewhere in the Boreal there is an alba, *M.p.* 'Alba' and a rosea, *M.p.* 'Rosea'. There will be some with strange pellucid dots, others with pellucid slashes of white. They would run screaming into the horticultural world of the global garden. And finally the Cree would benefit.

A Virginia bluebell, *Mertensia virginica*, is similar to the Boreal bluebell, *Mertensia paniculata* var. *borealis*. The Boreal bluebell should be developed as a garden beauty.

127

Bluebells

Monarda

BERGAMOT

Labiatae

Bergamot, *Monarda,* is a medicinal plant that spring-cleans the lungs.

128

Bergamot

The life of a plant, any plant, is complex. But when the plant reached into the ancient forests and lived with the kings of time and now remains today, then there is something very unusual about its life indeed.

Bergamot, *Monarda,* was the cause of a battle. It took place in the innocent waters of Boston Harbor. The descendants of the Pilgrims wanted their tea. It was the only luxury left to them by a religion of sackcloth and ashes. The British wanted to tax that tea, an idea not unusual to this day. The Bostonians got mad. Their rage boiled over. They dumped the tea into the harbor and in a snit began to drink the native brew, Oswego tea. They have continued. Nowadays it is refined. Earl Grey tea is served from silverware on pie-topped tables into porcelain cups. Occasionally, if the event deserves it, a sugared violet is floated on that tea.

Bergamot is also a Boreal plant. The best still live there. It is an aristocrat of the plant world and, like Jesus, has twelve disciples. These twelve species still live on the continent and are probably best represented by the Boreal bergamot, *Monarda fistulosa.* This is a purple-blue flower that arises from a handful of surface roots, with square translucent stems to head into a series of powder puffs that keep flowering in a nice kind of genteel, ragged way. The Cree have nodded their heads to it, calling the bergamot cow-pleasant-tasting-plant.

Bergamot can help the entire human race. As a member of the medicinal *Labiatae* family, it is filled with interesting volatile oils. The oils in bergamot are quite similar to those of the wild bergamot orange of Spain. A few of these oils act as dispersing agents. That is, they help the other and more medicinal oils travel and stay suspended in the air. These suspended oils travel into the lungs and other open areas of the body. They penetrate in small amounts. One of the most powerful is thymol. It has antiseptic and antifungal properties, cleansing the inner surfaces of the lung and nasal airways. These bergamot oils also act as a bronchodilator, cleansing out the lower regions of the lung. They are a dab hand at clearing out pollution products.

One product in particular is small. It is a killer. This pollution is less than 2.5 microns in size and is able to lodge in the deep regions of the lungs. These tiny things carry hitchhikers like heavy metals and pesticides. These multiply the damage of simple pollution, killing faster in the form of circulatory collapse. But bergamot can help to reverse this lodgement. The same bergamot that lived with the kings of time will arise to help us once again.

Nuphar

YELLOW WATERLILY

Nymphaeaceae

Interesting. Sex and the waterlily have been together for a long, long time. The name *Nuphar lutea* comes from the Arabic, *nilo-ufar,* and the Sanskrit, *nilotpala,* but these names are for the infamous blue waterlily of ancient Nile in Egypt. The frescos of the pharaohs were lavished with these blue flowers, symbols of erotic behavior, for this little sky-blue water flower of the Nile was an important aphrodisiac. Such items may well be necessary when brother marries sister in the long family line of the sun god.

The Boreal circumpolar forest and wetlands have their water lilies, too. They are butter yellow. The primitively simple flower forms itself into six sepals and holds its head above the muddy waters with a petiole that is elastic. The yellow, water-resistant stamens stand guard for the bulbous female fort in the dead center. The flowers, the young red leaves turning green with age, all come from a semifloating root as thick as a man's arm and just as strong.

True to form, the yellow waterlily of the Boreal is packed with surprises. They arise, as in all species of the margins, to greet us with a complexity of chemistry. It comes as a given that the quinolizidine and sesquiterpene alkaloids play the same or similar sexual tricks as the blue diva of the Nile.

The rhizome tissue shows an extraordinary range of unusual antibiotic and antifungal activity, especially with vectors to stop or stall the vegetative growth of yeasts, be they on the waterlily or the body of man. The tuber shows strong estrogenic activity and a strange enzyme that inhibits the synthesis of the prostaglandin group of chemicals, all of which are vital to the sexual life of man or woman, including childbearing and birthing. These body biochemicals reach deeply into the sexual life of mankind and act as a switching mechanism. For all the science of sexuality, this little yellow waterlily of the Boreal could streamline the sexual science of medicine with the mixture floating in its vacuoled, rough-and-ready tuber.

The mature tubers of the yellow waterlily were harvested after the flowers had finished blooming and were sliced and air-dried in the fall. This sun-dried medicine was used to treat arthritis, joint pains, swollen limbs, and diabetic ulcers of the skin, among a host of other ailments. It was also used as a cardiotonic. What other great medicines could our Boreal peoples bring to the world?

Waterlilies, *Nuphar,* are used to treat arthritis.

129

Yellow Waterlily

Polygonum

BISTORT

Polygonaceae

Bistort, *Polygonum*, is a Boreal food.

130

Bistort

Bistort is just one of a potential 80,000 species of plants that could feed the planet. Nowadays the food market is steaming with a love affair of just over twenty.

The leaf of bistort, *Polygonum viviparum*, is in the pharmacopeias of Switzerland, Russia, and France. The Slave nation of the Boreal forest have had more than a nodding acquaintance with the bistort over the millennia. They have enjoyed bistort, or *Polygonum viviparum*, with fried meats or fish.

They wash and slice the small dark rhizomes together with the odd smaller root nodule. They stick the whole lot into a sizzling frying pan and enjoy a special treat of the Boreal north that is packed with potassium like a banana. Bananas, like bistort, are good for filtration in the kidneys.

Bistort, *Polygonum*, is a tribe of some 150 species in the middle of which sit the ever popular rhubarb, spinach, chard, and kasha or buckwheat cereal, all darlings of the north. Bistort can also be found on the rosary of the Sakhalin Islands. Here the plant reaches extraordinary proportions and climbs to a 4 m (12 ft.) perennial. This perennial, with its tiny green flowers, is used as a rough forage crop, harvested in the fall, to which chickens and other avian life are truly addicted. This jolly green giant is called the Sakhalin bistort, *P. sachalinense*.

The bistorts have an unusual capacity as growing plants to concentrate important minerals from the subsoil. These are then to be found in the leaves in a water-soluble form much like iron in spinach. As the bistorts mature they produce complex tannic acids. These, together with an increase of oxalic acid, makes all the mature leaves bitter tasting and slightly toxic to some people. However, it is the mixture of tannic and oxalic acids, together with vitamin C and their complex starches, that gives the plants their medical capacity, especially for stomach problems and indigestion.

The aboriginal peoples of the south used *P. pennsylvanicum* in the treatment of horse colic, and another bistort, *P. persicaria*, as a plant rub to keep flies and insects from bothering horses during the summer months.

Bistort as *Polygonum punctatum*, water smartweed, was used with butternut, *Juglans cinerea*, and two other plant natives, the carrion flower, *Smilax herbacea*, and buttercup, *Ranunculus abortivus*, to increase blood flow in the brain during menses. Any female who has taken a math exam in the dire times of menses would be grateful for such medical relief bistort could offer.

Polypodium

COMMON POLYPODY FERN

Polypodiaceae

News flash. Maybe the ferns changed the world. Once upon an ancient time ferns occupied every corner of the planet. They made chemical warfare, too. They imposed themselves on the lowly life in a dictatorship of weapons. In their arsenal was a tasty treat. It was a sweet, hundreds of times sweeter than sugar. But it packed a genetic punch and radically changed the local plant DNA by fiddling with the spindle fibers of a reproducing cell, a rigid road to species change, leading into the realms of biodiversity.

In those wily times the climate was different. It was hot. It was warm and moist. The atmosphere was different, too. A sour carbon dioxide rode the air. And oxygen barely made a mark. The families of ferns formed an incredible tapestry, a flush of monstrous trees with strap-shaped shiny leaves fitted out to shed the rains. Glistening aerial roots like hang gliders and giant hairy rhizomes rose out of the earth like serpents as big as a man's arm and equally strong. These put forth the pinnules of leaves storing in their underbellies the housing of indusia. These were nurseries that shot their truculent spores out into an alien world.

Times have changed. The ferns have been slapped back into place by the bullying of gases. The atmosphere has flipped sides. The carbon ride has gone and oxygen rules the day. One family of ferns, the *Polypodiaceae*, can still be found hiding around rocks from the south to the north. The sweet steroid saponins are there, too. In Peru the *Polypodium* is called the medical fern or *calahuale*. In the Boreal north it is the fern that fixes tuberculosis. Somewhere in between the *Polypodium* species cure cholera.

Today, the Cree of the Boreal north crush the fresh rhizomes of common rock polypodium, *P. virginianum*, in the manage-

The common polypody fern, *Polypodium*, is used in aboriginal medicines worldwide.

ment of chest infections, coughs, and bronchial catarrh. More concentrated doses were used as a laxative and for the elimination of tapeworms from the body. The fern was also used to increase appetite.

One sweet steroid saponin of the polypody is called osladin. It is a true agent for change. It causes genetic mutation. In a planet saturated with osladins, change came as biodiversity. We see those changes today in the cones of the conifers, the display of our tropical forests, and in the ultimate coronation, the forest crown of the Boreal north.

131

Common Polypody Fern

Ribes

CURRANT

Saxifragaceae

The Arabs named the currant. They noticed that the leaves of the currant were like a small version of a rhubarb leaf, so they called the currant *rībās*. This got corrupted to *Ribes*, an umbrella under which the currants and the gooseberries live in the family *Saxifragaceae*.

The Boreal landscape is awash with currants. They can be seen in August when these little shrubs produce their great abundance of fruit. There are red currants, *Ribes glandulosum* and *R. triste*. *R. triste* is also found in the Siberian Boreal. There are black currants, *R. hudsonianum* and *R. americanum*. These differ little from the famous *R. nigrum*, the northern European species. There is also the extraordinarily fragrant buffalo currant, *R. odoratum* 'Crandall', that pushes up into the south of the Boreal. This fruit became a favorite of the pioneers, who bottled, canned, sauced, and hoarded it in every possible way.

All parts of the red and black currants are fragrant. Every leaf, stem, branch, and even the roots produce a strangely bitter smell that is found in the taste of the fruit.

To the peoples of the Boreal north the currant is an important bush food, one that maintains health. This opinion is carried over into the remainder of the Boreal world of Europe and Asia, where a taste for white, black, and red currants traveled and became mixed with European ethnic cuisines.

The aboriginal medicines of the Boreal currant all have a similar function, that of the protection of the capillary tissue of the female patient either in childbirth or as an induction for menstruation or to prevent miscarriage. The fruit has also been used universally either to prevent or to treat colds and flus.

The Cree used their frog berry, the wild red currant, *R. triste*, as a stem decoction to prevent blood clotting after birth. They also used the same species to bring on menstruation. In this case the mature dark stems were scraped. These scrapings were air-dried. They were then boiled in a decoction that was filtered and drunk.

The southern nations used the black currant for divination. A twelve-inch scraping was taken from a vernal twig. This was put into three liters (three quarts) of water and boiled down to one. Silver was put under the pillow together with a personal item of one who was sick. The decoction was drunk near sleep. And the subsequent dream would reveal all.

Rosa

PRICKLY ROSE

Rosaceae

A rose by any other name is the prickly rose of the Boreal north, *Rosa acicularis*. This little rose bush with its weak, straight thorns is at the mercy of the elements. In rainy seasons the spring canes can grow to taller spires of flowers, and in dry seasons these ruby canes will sulk but produce the same abundance of sweet-smelling open roses that are the pride of the northern world. The *Rosa acicularis* is seen throughout the circumpolar Boreal area in plentiful amounts. A sprinkling is seen, too, tracking the mountains and rougher terrain down into New Mexico.

The wild, single, delicious pink rose presents an open parabola to wasps, bees, and flying insects. In this parabolic form an insect with wings can maneuver its flight in many irregular circles and still feed. These acrobatics cannot be performed in the double flowering rose. The insects come in the spring as the male anthers display their yellow pollen to the public. A lactone aerosol is presented as an invitation to feed. The insects are mad for this golden food and bear it with banners of leg loads back to their propolis-lined homes.

The prickly rose is a medicine and a food. Both the petals and the hips are edible. The hips redden and become sweeter with the first touch of frost that leads to a greater sugar change in the hips' metabolism. This is when the hip is captured for food. The rose hip is high in vitamin C, but it has another complex of water-soluble vitamins B, E, and K that makes the food unusual. A tisane of the hips, a syrup, jelly, or jam, acts as an immediate tonic and reduces the tiredness of lethargy.

The rose has been used as a medicine since the beginning of time. Pliny of the Roman Empire certainly used it, as do the multinational cosmetic conglomerations of today.

The prickly rose has been used by the aboriginal women of the Boreal in the management of excessive bleeding that sometimes comes with menstruation. A decoction was drunk made of four mature, red-colored branches. This corrected the problem. This, too, treated snow blindness with eye drops. More mature branches or canes were chosen for this purpose, and these had a higher content of the vasopressor hesperidin.

The rose petals were used as a skin salve to reduce the swelling of stings. This is the underlying chemical basis of the oil of rose being used in cosmetics, and the prickly rose of the north is the best.

The single rose of the Boreal north, *Rosa acicularis*, feeds beneficial insects.

133

Prickly Rose

Rubus

RASPBERRY

Rosaceae

The common red raspberry, *Rubus idaeus*, is a cardiotonic.

An elder of the Boreal made an astute observation when he noted that as his peoples stopped eating bush food, or food from the wild, they no longer had their health. This observation is true for the remaining peoples of the world.

Edible plants like the red raspberry depend on being picked for the survival of their species. Picking is the dispersal of seeds, and seeds are that genetic lifeboat for survival. They must travel away from the mother plant to grow. A biochemical harmony arose over the millennia between the picker and the picked. This harmony benefits the picker, who is pulled back and back again to pick. The benefit is biochemical. In the case of the red raspberry and the 250 species of the *Rubus* tribe the biochemical is ellagic acid, a naturally occurring chemotherapeutic agent that stands guard on the picker's ultimate health and well-being.

Raspberries are the last ritualistic seasonal gathering of wild food. This happens in the Nordic countries and in the circumpolar Boreal, where death of a tree means life to a raspberry. The raspberry is picked for the food value in its berries and its medicine. The medicine is found in the leaves and in the primocane, that is, the new cane that does not bear flowers or fruit. It is found in the floricane, too, which bears flowers and fruit. The floricane quite often bears different leaves from the primocane, and the quality of the epidermal tissue is different and less glandular.

Raspberries and their medicinal canes have been used since time immemorial. The pips come up in the oldest archeological digs worldwide. The raspberry, *R. moluccanus,* is common to the Malay Peninsula and India. The Mauritius raspberry, which is native to eastern Asia and is widely grown in the tropics, holds legendary medicines. Of all the raspberries the common red raspberry, *R. idaeus,* is the most well known. This has modified itself into a more glandular being for the Boreal. It holds the greatest medicinal load of all of the red raspberries.

The primary biochemicals of the raspberry are fragarine complexes, gallotannins and ellagitannins, vitamin C, flavonoids, citric acid, and the catalyst zinc, among other known and unknown substances. These work as uterine stimulants and relaxants, speeding parturition. They are helpful in regulating menstruation, diarrhea, amoebic dysentery, and for the reduction of vomiting. They also serve as circulatory and cardiotonics.

The simple condition of sloth is treated with a decoction of raspberry, *R. idaeus,* and ginseng, *Panax quinquefolia.* The raspberry bumps the sex organs, and the panacea of the ginseng jump starts the rest of the body . . . so the motor of life is rolling . . .

Sarracenia

PITCHER PLANT

Sarraceniaceae

Turn your back on some plants and they will do unspeakable things. The pitcher plant, *Sarracenia purpurea,* is one such Boreal beast that dabbles in death. This plant likes its meat fresh, twitching a little, and then settles down to digestion in serene silence.

The pitcher plant is a product of its own habitat. The plant likes bogs and wet places where the availability of nitrogen is low. This essential element is needed for general housekeeping within the cells. It is needed in increased amounts when the plant wants to divide or reproduce. Then the need is at a fever pitch.

Over time the pitcher plant, like its seven other siblings, developed a peculiar cunning. They curled their leaves around and sealed them to form a pitcher or vessel that caught water. This was innocent enough because many plants such as the monocotyledons do this in the axils of their leaves. But the pitcher plant went a few chilling steps further.

The leaves developed an outside curl that became outlined with borders of blood-red advertising. The veins became the same deep scarlet of pulmonary clotting. Then the leaves grew an inside palisade of glandular hairs above the pool.

Hovering over the pool on a long and delicate pink stalk is a flower of magnificent proportions, arising like a red lantern in a sea of green. The nectaries of this red-light district produce sexual overtones, in betulin, amyrin, and lupeol, all progenitors of sexuality. The insects come willingly as clients into this striptease of illuminated red glamour.

They drop in for a drink. The pool is at their feet. They buzz down for a dip. They are a little careless because upstairs in flower heaven the going was so good. But they cannot get out.

The pitcher plant, *Sarracenia purpurea,* likes its meat fresh.

The ultimate terror is in their path: death and destruction by digestion. As the last nitrogen atom is sucked up through the circulation of the plant, it relaxes into growth and further splendor. The queen of carnivores respires gently in wait for more prey.

The pitcher plant is considered by the medicine men and women of the North American aboriginal world to hold the most powerful medicines. The complex medicine of this plant has been used for witchcraft, lovesickness, fevers, kidney disease, and migraines. It has been used as an oxytoxic childbirth medicine and to regulate menses. It is a cardio and capillary tonic and it is thought to have antileukemia properties.

The beast of the Boreal may well turn into a beauty after all.

Pitcher Plant

Solidago

GOLDENROD

Compositae

Goldenrod, *Solidago,* was a flagship medicine for the Arabs during the Crusades.

Goldenrod

The flowering species known as goldenrod, *Solidago,* makes it way all over the world. There are dry fields in the Norway sun filled with golden flowers. It is in the Azores, northern Africa, and Asia. But the goldenrod with its special need for the driving heat of a midday sun has made the North American continent its favorite home. Here on the flat, dry face of the "turtle" the *Solidago* species can even be found within a hair's breadth of the Atlantic Ocean, sitting in salty sand. The goldenrod makes its remarkable way into the Boreal, too. There, it is, as it is everywhere, a medicinal species *par excellence.*

In the meadowlands and forest edges of the Boreal, the *Solidago* turns a face to the last solar trick of the sun. In Canada, the Canadian *Solidago, S. canadensis,* turns, and in the Russian and European circumpolar regions it is *S. virgaurea.* The heat fires the nectaries of the ovaries to produce their complex polysaccharides for the insect world. This sweet solution is mixed with many medical calling cards, one of which is a glucocorticoid agent, and another is an oil that is seductively volatile and is a last call to pollination. This is never missed. The flying insects come pick up their flavones, too. This is to get them and their kin through the lifeless winter months.

The global history of the goldenrod is interesting. After the slaughter of the first and second Crusades, the Arabs mopped up their wounds with goldenrod. This specialized knowledge came out of one nomadic tribe, the Saracens, who wandered between Syria and Arabia. This medicine caught on like wildfire with the Arabic peoples, so much so that the nearby Italians called the goldenrod the "pagan herb," *erba pagana.* This was a Christian ivory tower reference to the great divide of religion in which they lived. The Germans on the other hand called goldenrod *Consolida saracenia,* a herb they carefully noted would make one whole or healthy again.

The aboriginal peoples of the Boreal used a decoction of the mature stem and leaves to treat both kidney and bladder problems. These are similar to the uses of *Solidago* all over the remainder of the world for jaundice, biliousness, and various liver disorders. Another, *S. rugosa,* was used for sunstroke and dizzy spells. A mixture of goldenrod and ginseng, *Panax quinquefolia,* was used for the lovelorn to kill their agony of love.

The goldenrod honey of North America is a healing honey. It is filled with the real gold of autumn's gilded pastures.

Sphagnum

PEAT MOSS

Sphagnaceae

The name tells it all. Peat moss is a moss that makes peat. It makes coal and diamonds, too, if the pressure is correct. Peat moss is partly responsible for the vast undersea methyl hydrate reserves that are sheer liquid gold waiting to be plundered by an energy-hungry world.

Peat moss is a member of the scraggy-necked family called *Sphagnaceae*. Most peat moss is a mix of *Sphagnum* species. *Sphagnum* pulled itself out of the oceans' seaweeds a couple of hundred million years ago and made its life on land. This moss did it by turning a sexual trick. *Sphagnum* takes part in the usual marriage of male to female, egg and sperm. But *Sphagnum* drags its feet and the fertilized egg doesn't leave home as a spoiled sphagnum brat. It stays on in the house. In this armchair a changeling arises. This arrangement of family is a throwback to its oceanic life.

The changeling sphagnum brat begins a life of lust as a sporophyte generation. A head is formed called a calyptra. This head is fine-tuned to atmospheric moisture levels. When it is paper dry, and then gets a little drop of moisture, the head snaps open to reveal a grinning set of peristone teeth. These vegetable dentures open rapidly to release the gang of brats in the multimillions. The catapult action of this opening is ballistic. The gang of brats is set free as spores to roam the world. It was something they did once upon a time in a massive global colonization when the planet was hot and wet. Then they ruled as giants.

Now these changelings still regulate the world. They do it with the aid of a modified internal tissue called hyaline. Hyaline cells of *Sphagnum* can hold a thousand times their own volume of water. Like a sponge, they don't change shape. They become heavier and denser. This moss, *Sphagnum*, takes up the water in bogs, fens, lakes, wetlands, and riparian areas that yawn across the planet. They especially maintain the humidity of the vast circumpolar wetlands of the Boreal upon which insect, bird, and mammal life depends. This circle of polar moisture resides beside the arctic cold. The moisture is held in place by the condensing cold of the north. The sphagnum holds this moisture in its hyaline tissue and multiplies it into a dripping spongelike consistency. This wet skin lying on an almost sterile arctic soil provides the humidity that grinds the medicines out of the Boreal forest, a forest that, by rights, should never be there, because of too little sunlight. It is there nonetheless, to pump medicinal aerosols into the north winds that blow and help to protect our entire world. And keep the Boreal babies dry.

Sphagnum moss carpets the Boreal in a protective layer.

137

Peat Moss

Typha

CATTAIL

Typhaceae

The cattail, *Typha latifolia*, is a cosmopolitan genius. This is a wetland species. It lives in Europe, in Asia, and in North America. It travels in the waterways of the Boreal north.

The cattail is a monocotyledon. This means that this perennial plant is related to the grasses, having a similar anatomy of the root and stem. The cattail lives in water as a thick trailing rhizome, about the size of a wrist. From this underwater mass grows a green sheath of leaves that curl around like corn. The leaves have a system of parallel veins typical of the grasses. From the center of the sheath a terminal flowering spike shoots up a good 2 m (6 ft.), way above the water line. This spike has at its tip the typical brown bulrush flower, with its female parts below and the softer male flowers above, each with two to five united stamens. As it grows, the rhizome underwater produces a series of fat, emerging, semidormant shoots. These add considerably to the overall bulk of the rhizome.

The cattail is a traditional vegetable of the aboriginal peoples all over the world. Every part of this plant is edible. In Siberia, it is called Cossack asparagus. In the Canadian Boreal forest, it is called grass fat by the Chipewyan. The young shoots in the spring can be eaten like asparagus. The inner white stem can also be eaten. The mature rhizomes can be peeled. They can be either roasted or fried, the regions with the dormant shoots being especially good. The seed heads can be dried and used as a kapoklike filling.

Cattails are of enormous importance to the wetlands of the global garden. They regulate water flow and maintain the integrity of the marshlands. They are nurse species for many other oxygenating water plants. They help to oxygenate the surrounding water. They shade it. They reduce evaporation. They provide food for wildlife and shelter too. They adsorb toxic divalent cation pollution. The leaves are used for sunbathing, increasing the vitamin D ratios of insects, birds, and amphibians, thereby improving their fertility.

The cattail is the green building block that makes the millions of square kilometers of wetland work in the global garden. This is its genius. This is its design.

Umbilicaria

ROCK TRIPE

Umbilicariaceae

Tripe is one of those foods that other people eat. Tripe is the stomach of a ruminant, usually a cow, dead of course. This tissue is washed several times in running water. A white floppy piece of meat complete with its worried-looking lining is delivered as a soup with an onion and a small sneeze of black pepper as a serving detail to the delighted.

Tripe, a vegetable tripe, with similar facial details occurs on rocks. Many of them are to be found in the Boreal and others splashed across rocks in the remainder of the world. Like the animal variety, though, rock tripe is edible also. It, too, requires washing with water and needs a base of some kind like hardwood ashes or bicarbonate of soda to neutralize the vulpinic acid, which quite often can be toxic and at best an irritant to the gastrointestinal tract.

Rock tripe of the *Umbilicaria* species, like peppered rock tripe, *U. deusta*, blistered rock tripe, *U. hyperborea*, and plated rock tripe, *U. muehlenbergii*, have been eaten by the hardy individuals of the Huron, the Naskapi, the Chipewyan, and of course the Cree. Some of these folk enjoy a stomach ice cream, a flavor as yet uninvented by the gelato industry. This ice cream involves harvesting the partially digested tripe lichen mixture from the stomachs of slain caribou. This steaming matter is added to raw fish eggs and gently stirred to form one of the northern national dishes enjoyed by many.

The Japanese are not slow in their appreciation of rock tripe, *U. esculenta*, having stretched their species into legendary food. Their scrapings are beheld in rich restaurants by the name of *iwatake* on the specialty menus. Their tripe is undulated into soups or salads. Because it has the earthy taste of mushrooms

just at the point of composting, *iwatake* translates into the most desirable name of rock mushroom.

The monastery system of Siberia was not excluded from the effects of lichens either. One particularly boisterous group of monks gave in and added a little lichen for bitters to fire up their alcohol, raising their spirits closer to God.

The circumpolar rock tripe is an emerging food. It changes the nature of rock. It provides shelter for insects and nesting for birds. It is of itself a daunting success. A marriage between fungi and algae. A trailblazer for biodiversity.

Rock tripe, *Umbilicaria*, makes a Boreal ice cream.

Rock Tripe

Vaccinium

BLUEBERRY

Ericaceae

One hundred nights of subzero temperatures bring the blueberry to our tables. This fascinating member of the heath, *Ericaceae,* family is a favorite fruit of the Western world.

The blueberry, *Vaccinium angustifolium,* together with its wild brothers of the blueberry barrens, is not a new fruit. It had its own name in the Middle Ages, when it was known as *Mora agrestis,* presenting itself as a respectable medicinal plant. The blueberry was known well before that, for it was carefully named in the ancient writings of the past.

The Boreal north has known about the plant and its berries for millennia. The aboriginal nations place great importance on their bush food, which the medicine men and women firmly believed was the source of their health. They picked the ripe berries cheek to jowl with black bears. They dried them over a cool fire for winter storage. And sometimes they stored them fresh in lard to tempt the palate and remember the summer months.

There is something decidedly odd about the blueberry. The plant loves the acid soils, no doubt, and grows well in them, making the powder blue fruit for which the skinny plant is famous. But the roots have no root hairs. They manage their underground life dealing with mycorrhizal growth in a mysterious manner. From its lucky dip of growth the plant manages to store some strange elements in the fruit. These are chromium, zinc, iron, copper, magnesium, and molybdenum. These are placed into a form that is extremely healthy to eat. All these elements make up the co-enzyme catalytic factors that fire up all metabolic pathways in the moving man, beast, or bird.

Blueberries hike up the iron in hemoglobin. They are high in vitamin C and quercetin, the universal capillary protector. They therefore clear the skin and beautify the face. The ripe blueberry is antidiabetic and somehow helps with hypoglycemia.

The future looks good for the blueberry, too. They have been cross-pollinated and hybridized for over one hundred years, and bigger and better blueberries are coming into the markets every day. All of these blueberries taste divine, but none of them is as fine on the tongue as the wild blueberry of the Boreal north.

Vaccinium

BOG CRANBERRY

Ericaceae

A new juice has been rolled into the food industry. It is the juice of the cranberry either alone or in its various combinations, as a fruit cocktail. The juice is a deep red color and has a bitter taste and a high acid content. The demand for it is great and growers are seriously getting into gear to increase its production.

The cranberry has been known for a very long time. It is mentioned in ancient writings, and the name itself stems from the Low German word for crane, *craan*. All over the global garden these wading birds love this little red juicy berry, particularly after a frost.

The cranberry, *Vaccinium oxycoccos,* is a heath species loving extremely acidic soils. The creeping evergreen plant is found in colossal numbers, particularly in maritime areas tracking the coastline of the Boreal shores of the North American continent into Iceland and most of Eurasia.

The Canadian maritimers celebrated a homespun cranberry day in the past. The children were sent off to inspect the state of the fruit along the capes of the seashore. Then the stage was set for a communal picking. A full year's supply was brought into the dugout basements as pickles, relishes, jellies, jams, juice, and canned whole berries.

There are 150 cranberry species in the global garden. A particular favorite of Norway, Sweden, and Finland is the mountain cranberry, *V. vitis-idaea* ssp. *minus.* This is the beloved lingonberry added to so many ethnic dishes. There is also the sweetly fragrant Kamchatka bilberry, *V. praestans,* of the Kuril Islands.

The cranberry is a health food. It has been used as a urinary antiseptic. Cranberry juice helps to flush the urinary tract free of

E. coli. These bacteria are on the increase because of population growth and the abuse of antibiotics in industrial farming and fishing. In addition, estrogen-mimicking toxins alter the body's natural immune system, lowering resistance to *E. coli.*

Cranberry juice contains condensed tannins and a specially modified form of glucose called vacciniin. These intercept the ability of the *E. coli* to attach itself to the cells of the urinary tract, keeping these pathogens away.

The juice of cranberries wards off *E. coli.*

141

Bog Cranberry

References

Beresford-Kroeger, Diana B. *Arboretum America: A Philosophy of the Forest.* Ann Arbor: University of Michigan Press, 2003.

Beresford-Kroeger, Diana B. *Bioplanning a North Temperate Garden.* Kingston, Ont.: Quarry Press, 1999.

Beresford-Kroeger, Diana B. "Canada Eh! Show Your Colours in the Garden with 10 Easy-to-Grow Native Perennials." *Nature Canada,* spring 2001, 32–35.

Beresford-Kroeger, Diana B. "Frost Resistance and the Importance of the Gibberellin Complex in the Growth of Trees." Masters thesis, University College Cork, Cork, Ireland, 1965.

Beresford-Kroeger, Diana B. *A Garden for Life.* Ann Arbor: University of Michigan Press, 2004.

Beresford-Kroeger, Diana B. "The Ideas of Diana Beresford-Kroeger." *Ideas,* CBC, March 21, 2005, P.O. Box 500 Station A, Toronto, Canada, M5W 1E6.

Beresford-Kroeger, Diana B. "Just What the Doctor Ordered." *Nature Canada,* spring 2001, 16–17.

Beresford-Kroeger, Diana B. "King of the Forest." *Nature Canada,* spring 2000, 16–19.

Beresford-Kroeger, Diana B. "Preserving the Butternut Tree." *Eco Farm and Garden,* spring 2003, 44–47.

Beresford-Kroeger, Diana B. "The Promiscuous Plant." *Nature Canada,* autumn 1998, 24–25.

Beresford-Kroeger, Diana B. "A Summer Beauty." *Nature Canada,* winter 1999, 18–19.

Beresford-Kroeger, Diana B. "Synthesis, Metabolism and Importance of Indole Glycosides and Serotonin Hormones in the Plant Kingdom." Doctoral research, unpublished, Ottawa, Ont., 1970.

Beresford-Kroeger, Diana B. "Three Plants in One." *Nature Canada,* spring 1999, 31–32.

Boon, Heather, and Michael Smith. *The Botanical Pharmacy.* Kingston, Ont.: Quarry Press, 1999.

Borror, Donald J., and Richard E. White. *A Field Guide to the Insects of America North of Mexico.* Boston: Houghton Mifflin, 1970.

Bower, B. "Evolution's Child: Fossil Puts Youthful Twist on Lucy's Kind." *Science News,* September 2006, 195.

Brodo, Irwin M., Sylvia Duran Sharnoff, and Stephen Sharnoff. *Lichens of North America.* New Haven: Yale University Press, 2001.

Budavari, S. *The Merck Index: An Encyclopedia of Chemicals, Drugs, and Biologicals.* 11th ed. Rahway, N.J.: Merck, 1989.

Casselman, Bill. *Canadian Garden Words.* Toronto: Little, Brown, 1997.

Clausen, Ruth Rogers, and Nicholas H. Ekstrom. *Perennials for American Gardens.* New York: Random House, 1989.

Cody, J. *Ferns of the Ottawa District.* Ottawa: Canada Department of Agriculture, 1956.

Collingwood, G. H., and Warren D. Bush. *Knowing Your Trees.* Washington, D.C.: American Forestry Association, 1974.

Cormack, R. G. H. *Wild Flowers of Alberta.* Edmonton: Queen's Printers, 1967.

Davies, Karl M., Jr. *Some Ecological Aspects of Northeastern American Indian Agroforestry Practises.* Northern Nut Growers' Association, Annual Report 85, 1994, 25–39.

Densmore, F. *Indian Use of Wild Plants for Crafts, Food, Medicine, and Charms.* Ohsweken: Iroqrafts, 1993.

Farris, Cecil W. *The Hazel Tree.* East Lansing, Mich.: Northern Nut Growers Association, Michigan State University, 2000.

Fell, Barry. *Bronze Age America.* Toronto: Little, Brown, 1982.

Flint, Harrison L. *Landscape Plants for Eastern North America.* New York: John Wiley and Sons, 1983.

Fox, Katsitsionni, and Margaret George. *Traditional Medicines.* Cornwall: Mohawk Council of Akwesasne, 1998.

Hamilton, J. W. "Arsenic Pollution Disrupts Hormones." *Science News,* March 2001, 164.

Herity, Michael, and George Eogan. *Ireland in Prehistory.* New York: Routledge, 1996.

Herrick, James W. *Iroquois Medical Botany.* Syracuse: Syracuse University Press, 1995.

Hillier, Harold. *The Hillier Manual of Trees and Shrubs*. Newton Abbot: David and Charles Redwood, 1992.

Hosie, R. C. *Native Trees of Canada*. Ottawa: Department of Fisheries and Forestry, 1969.

Howes, F. N. *Nuts: Their Production and Everyday Use*. London: Faber and Faber, 1948.

Howes, F. N. *Plants and Beekeeping*. London: Faber and Faber, 1979.

Jaynes, Richard A. *Nut Tree Culture in North America*. Hamden, Conn.: Northern Nut Growers' Association, 1979.

Klots, Alexander B. *A Field Guide to Butterflies of North America, East of the Great Plains*. Boston: Houghton Mifflin, 1951.

Krieger, Louis C. *The Mushroom Handbook*. New York: Dover, 1967.

Krochmal, Arnold, and Connie Krochmal. *The Complete Illustrated Book of Dyes from Natural Sources*. New York: Doubleday, 1974.

Lee, Robert Edward. *Phycology*. 2d ed. Cambridge: Cambridge University Press, 1995.

Layberry, Ross A., Peter W. Hall, and J. Donald Lafontaine. *The Butterflies of Canada*. Toronto. University of Toronto Press, 1998.

Lellinger, David B. *A Field Manual of Ferns and Fern Allies of the United States and Canada*. Washington, D.C.: Smithsonian Institution Press, 1985.

Lewis, Walter H., and Memory P. F. Elvin-Lewis. *Medical Botany: Plants Affecting Man's Health*. Toronto: John Wiley and Sons, 1979.

Liberty Hyde Bailey Hortorium. *Hortus Third: A Concise Dictionary of Plants Cultivated in the Unites States and Canada*. New York: Macmillan, 1976.

Little, Elbert L. *Trees*. New York: Alfred A. Knopf, 1980.

Marles, Robin J., Christina Clavelle, Leslie Monteleone, Natalie Tays, and Donna Burns. *Aboriginal Plant Use in Canada's Northwest Boreal Forest*. Vancouver: University of British Columbia Press, 2000.

Megan, Ruth, and Vincent Megan. *Celtic Art*. London: Thames and Hudson, 1999.

Mullins, E. J., and T. S. McKnight. *Canadian Woods: Their Properties and Uses*. Toronto: University of Toronto Press, 1981.

Myers, Norman. *Gaia: An Atlas of Planet Management*. New York: Doubleday, 1984.

Perkins, S. "Changes in the Air: Variations in Atmospheric Oxygen Have Affected Evolution in Big Ways." *Science News,* December 2005, 395–96.

Perkins, S. "Mercury Rising: Natural Forest Wildfires Release Pollutant." *Science News,* August 2006, 134.

Perkins, S. "Northern Refuge: White Spruce Survived Last Ice Age in Alaska." *Science News,* August 2006, 84.

Peterson, Roger Tory, and Margaret McKenny. *A Field Guide to Wildflowers of Northeastern and North-central North America*. Boston: Houghton Mifflin, 1968.

Phillips, Roger, and Martyn Rix. *Perennials*. 2 vols. New York: Random House, 1991.

Pirone, P. P. *Tree Maintenance*. 6th ed. Oxford: Oxford University Press, 1988.

Rackham, Oliver. *The Illustrated History of the Countryside*. London: George Weidenfeld and Nicolson, 1994.

Raloff, Janet. "Aquatic Non-scents." *Science News,* January 2007, 59–60.

Rupp, Rebecca. *Red Oaks and Black Birches: The Science and Lore of Trees*. Pownal, Vt.: Storey Communications, 1995.

Schopmeyer, C. S. *Seeds of Woody Plants in the United States*. Washington, D.C.: Forest Service, U.S. Department of Agriculture, 1974.

Small, Ernest, and Paul M. Caitling. *Canadian Medical Crops*. Ottawa: National Research Council of Canada, 1999.

Smith, Russell J. *Tree Crops: A Permanent Agriculture*. New York: Devin-Adair, 1953.

Stuart, Malcolm. *The Encyclopedia of Herbs and Herbalism*. London: Orbix, 1979.

Index

Page numbers in italics refer to illustrations.

Index

Photography

Photographs for this text were taken with a Canon TL-QL 35mm SLR camera using the following Canon lenses: FD 70-210 mm zoom F/4, FL 50 mm F/1.8, and FD 28 mm F/2. For virtually all the photographs, Kodachrome 64 or Kodachrome 200 color transparency film was used. All photographs were made in natural light without enhancement.